Patrick Moore's Practical Astronomy Series

Other titles in this series

Telescopes and Techniques (2nd Edn.)
Chris Kitchin

The Art and Science of CCD Astronomy
David Ratledge (Ed.)

The Observer's Year
Patrick Moore

Seeing Stars
Chris Kitchin and Robert W. Forrest

Photo-guide to the Constellations
Chris Kitchin

The Sun in Eclipse
Michael Maunder and Patrick Moore

Software and Data for Practical Astronomers
David Ratledge

Amateur Telescope Making
Stephen F. Tonkin (Ed.)

Observing Meteors, Comets, Supernovae and
other Transient Phenomena
Neil Bone

Astronomical Equipment for Amateurs
Martin Mobberley

Transit: When Planets Cross the Sun
Michael Maunder and Patrick Moore

Practical Astrophotography
Jeffrey R. Charles

Observing the Moon
Peter T. Wlasuk

Deep-Sky Observing
Steven R. Coe

AstroFAQs
Stephen F. Tonkin

The Deep-Sky Observer's Year
Grant Privett and Paul Parsons

Field Guide to the Deep Sky Objects
Mike Inglis

Choosing and Using a Schmidt-Cassegrain
Telescope
Rod Mollise

Astronomy with Small Telescopes
Stephen F. Tonkin (Ed.)

Solar Observing Techniques
Chris Kitchin

Observing the Planets
Peter T. Wlasuk

Light Pollution
Bob Mizon

Using the Meade ETX
Mike Weasner

Practical Amateur Spectroscopy
Stephen F. Tonkin (Ed.)

More Small Astronomical Observatories
Patrick Moore (Ed.)

Observer's Guide to Stellar Evolution
Mike Inglis

How to Observe the Sun Safely
Lee Macdonald

Astronomer's Eyepiece Companion
Jess K. Gilmour

Observing Comets
Nick James and Gerald North

Observing Variable Stars
Gerry A. Good

Visual Astronomy in the Suburbs
Antony Cooke

Astronomy of the Milky Way: The Observer's
Guide to the Northern and Southern Milky Way
(2 volumes)
Mike Inglis

The NexStar User Guide
Michael W. Swanson

Observing Binary and Double Stars
Bob Argyle (Ed.)

Navigating the Night Sky
Guilherme de Almeida

The New Amateur Astronomer
Martin Mobberley

Care of Astronomical Telescopes and
Accessories
M. Barlow Pepin

Astronomy with a Home Computer
Neale Monks

Digital Astrophotography: The State of the Art
David Ratledge (Ed.)

How to Photograph the Moon and Planets with Your Digital Camera

Tony Buick

With 312 Figures

 Springer

Tony Buick
Orpington, Kent, UK

British Library Cataloguing in Publication Data
A catalogue record for this book is available from the British Library

Library of Congress Control Number: 2005927381

Patrick Moore's Practical Astronomy Series ISSN 1617-7185
eISBN: 1-84628-046-X
ISBN-10: 1-85233-990-X
ISBN-13: 978-1-85233-990-6

Printed in Singapore/KYO

9 8 7 6 5 4 3 2 1

Springer Science+Business Media
springer.com

To my two dear sons Chris and Tim for their support in everything always

Foreword

This is a remarkable book, unlike any other known to me. Tony Buick is not a professional astronomer; he is concerned with medical science. He does not aspire to own wildly expensive astronomical equipment – and yet with an ordinary digital camera and a very modest telescope he has produced lunar and solar photographs, which are fully equal to the best professional photographs available only a few decades ago.

His methods are straightforward enough, and what he has achieved here can be emulated by others, even allowing for the fact that Tony Buick has more than the average amount of practical skill. Follow his instructions, and you too will be able to produce stunning images which will not only give you pleasure, but may well turn out to be useful for research purposes.

This book concentrates upon photographs of the Moon, and to a lesser extent the Sun and planets, but the methods can well be extended to "deep space" objects such as nebulae and galaxies. There is a tremendous amount of scope. Tony Buick has shown the way, and what he has achieved will inspire many others to follow suit.

Sir Patrick Moore, CBE, FRS

Preface

This book is intended for the great number of amateur astronomers who wish to pursue the hobby of astrophotography but find it a daunting prospect to plough through books and/or otherwise gain information to translate such desires into the production of quality pictures.

Astronomy generates awe in us all. We watch the news showing spacecraft blasting off from their launch pads, amazing pictures from Mars and other planets, as well as eclipses of the Sun and Moon. Participation in such events seems too remote a possibility for many. And yet there is a way, that is so exciting, to capture astronomical images with only a small amount of effort and an even smaller amount of knowledge. Never mind the light pollution and tiny patch of sky available to you. You don't need a huge budget either.

In the following chapters a step-by-step guide is given to enable acquisition and construction of inexpensive equipment with which to facilitate the taking of beautiful photographs of space objects. No astronomical or optical physics knowledge is assumed or required save the small amount given where necessary.

Amateurs at all levels may find the speed and rapid accumulation of quality sky images useful for education and presentation purposes in addition to displaying those images as part of an absorbing interest.

In places, short astronomical introductions to the target objects are provided for beginners to peek into the vastness of space and time, to imagine the processes that formed them and to wonder at the strange cosmic bodies that exist.

It is hoped that enjoyment of the images obtained will lead to enthusiasm and continued development of this fascinating hobby.

Tony Buick
Orpington, Kent, UK

Acknowledgments

I am most grateful to Sir Patrick Moore for his kind encouragement and help to make possible production of this book. Also to my son Chris, without whom the task would have been so much more difficult – his computing, proofreading skills and ideas throughout have been invaluable – and my son Tim for computer enhancement of Jupiter and Saturn images and constant support for my project. I am indebted to the Orpington Astronomical Society members for their friendship, advice and enthusiasm. Thank you Jenny, Roy and Wendy for much support and constantly listening to my Moon ramblings and Gilbert Satterthwaite for his generous advice and encouragement.

Tony Buick
Orpington, Kent, UK

Contents

1   Introduction 1

2   Equipment 3

3   The Magic Ingredient 13

4   Method 21

5   The Universe and You 31

6   Targets 39

7   Our Moon 43

8   The Moon – First Glance 55

9   Regions of the Moon 63

10 Moon Features and Techniques 95

11 Lunar Events 167

12 Solar System Moons 197

13 The Planets 207

14 The Sun 221

15 Transits 231

16 And What Else? 247

17 In Conclusion 253

Appendix 255

Index 269

CHAPTER ONE

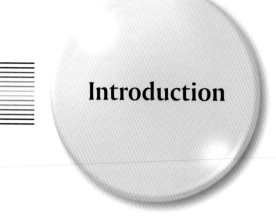

Introduction

You have a digital camera but do you have a telescope? Can you borrow one? Are you visiting someone who has one? Focus the telescope on something. Anything! The Moon, a distant telegraph pole, a boat on the horizon … Hold your digital camera very steadily over the eyepiece and take a picture. Your first telescopic image (figure 1.1) – and you didn't need any technical knowledge at all. You only required a digital camera, a cheap or borrowed telescope and a computer to store and print your images. Your hobby has already started or, for existing amateur astronomers, taken on a new dimension. This picture was taken with a very cheap camera and a small telescope. Some features of the Moon are clearly visible, such

Figure 1.1.

as the dark areas called maria ("seas") and the splendid bright rays from the crater Tycho.

I very much like the principle of "do first and learn later". It can be terminally off-putting to have to endure hours, weeks or many months of instruction or reading before producing your first image of the sky. Of course many mistakes will be made and many pictures will have to be erased but since you are transferring images onto a computer from your camera the expense of those mistakes will be as close as you can get to zero. (Purists please do not pick up on this illustrative statement.) Your pictures will be seen as soon as you can get to a computer – no waiting around for days to view disappointing results on film by which time your image opportunity has changed or even vanished.

Almost certainly your reaction to your first astronomical image will be *Wow*! Look at that detail! Did I do that? Then you are hooked. You will want to do more and know more and show your friends and colleagues with pride the sky images you have captured on camera. You will produce pictures ever more stunning even with your inexpensive starting equipment. It is extremely satisfying to produce the best possible pictures with the instruments you start with. Investigate all parameters on the telescope and the camera. Use your imagination, creativity and DIY (Do It Yourself) talents as a substitute for shelling out tons of money, which you can do later if desired to match your astronomical ambitions.

The following pages describe a journey for you to follow from never having produced a single astronomical image to the production of images that some experts will describe as stunning.

I recall one memorable conversation when showing off some early efforts.

Me: "Look at these pictures that I took last night."

Him: "Wow! They are stunning. I suppose you are in a low light-pollution area."

Me: "No! I have a street light right outside my patio viewing area."

Him: "I suppose you have good all-round vision of horizons."

Me: "No! I live on the side of a hill (east-blocked), with houses on each side of my patio and a streetlight at patio height and 7 metres to the west. I can only see a patch of zenith (overhead) sky."

Him: "I suppose you have expensive equipment."

Me: "No. I used my standard family digital camera and a broken second-hand telescope."

And so it went on. The message is to use whatever you have, can borrow or buy cheaply to begin with and use wherever you are.

It must be emphasized that *your* camera and telescope may well be a different make and model from those used as examples in this work but the principles are the same for you to transfer to *your* instruments.

A glossary is given in the appendix of technical or astronomical terms used in the text.

CHAPTER TWO

Equipment

Overview

Producing astronomical pictures won't cost you nothing. To avoid misinterpretation of that double negative – it will cost you something but the costs can be minimized to be within the reach of most pockets.

If you have only a telescope and digital camera with a back screen viewer then enjoy viewing your night's work. Very satisfying but all your pictures will have to be erased prior to the next session or sessions depending on how many pictures you take per night.

Maybe your camera does not have a screen viewer. Then the minimum required is a printer that can be directly linked to the camera to print selected or all images. Some cameras and printers provide a picture editing function to make simple changes before printing, such as cropping and sepia/monochrome choices. Also available are camera docking stations that allow viewing of all pictures taken on a TV or built-in screen and printing of images selected. Technology in this regard proceeds at a pace.

As the price of home computers continues to fall it is possible that you have one that you can use for your astronomy thus providing a major step forward in editing and presenting images. Camera plus telescope plus computer will give the most versatility, satisfaction and potential for showing off your super photos however they are obtained.

A good pair of binoculars, say 10 × 50, can be used with your camera for recording easy sky images but you will require some creativity and elementary DIY skills to achieve a sharp enough picture of which to be proud. See the Appendix for a suggested construction of a platform to hold camera and binoculars.

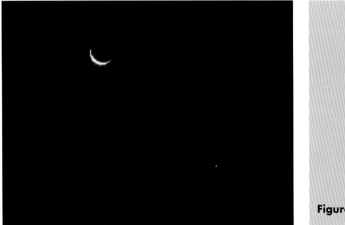

Figure 2.1.

The first really useful telescope with which to capture detailed sky images will have an aperture (diameter of lens or mirror) of around 60 or 70 mm. Features of the Moon and Jupiter's four largest moons will be clearly seen. The first glimpse of Saturn's rings is a breathtaking moment for any beginner. They really are there! Not just pictures in a book. And *you* can see them!

I suggest the optimum size of telescope for results and portability to be a 200 mm reflector (see later for suggestions). Here is where the stunning detail of the Moon and some planets can be seen. Even better – the stunning views can be transformed into stunning pictures.

What can you expect from each of these choices? Firstly, with a digital camera only (without telescope) the possibilities are limited but not zero (figures 2.1 and 2.2).

Both pictures were taken shortly before dawn with the camera resting on the car roof. Shutter delay was required, of course, to prevent camera shake, and camera zoom. The second picture clearly shows Earthshine where exposure time

Figure 2.2.

Figure 2.3.

has been increased through spot-meter focusing on a dark patch of sky and the image lightened on the computer. Other methods of increasing exposure are available and are discussed later. The small bright spot bottom right is Jupiter, which creates just a little more interest. Nice first pictures for your album. The pictures can be cropped from the camera and printed directly or loaded to a computer for more control over the final result.

Obtaining pictures (here's one – figure 2.3) through binoculars is trickier but a method is discussed in the appendix. The difficulties are being able to combine focusing, maintaining the target in view and keeping everything still whilst taking the picture.

Progression to a 60 or 70 mm telescope is when some serious astronomical observations can take place. The beautiful rings of Saturn are now just visible and the first sight of Moon details is mind-blowing. The rays of Tycho, the dark seas of solidified lava, or maria, and the features of craters are indicators of our fascinating natural satellite.

Figure 2.4

This book is particularly aimed at getting the most out of a digital camera connected to a 200 mm reflecting telescope and with the availability of a computer. The versatility of a modestly priced digital camera allows the collection of stunning astronomical images. With a clear and calm night pictures like this (figure 2.4), can be taken and printed in a matter of minutes (make that an hour).

Detailed Description

Camera

The advance of digital technology has made available a most sophisticated computerized camera at a price to enable anyone to buy one for around the same price as a film camera. Indeed digital cameras are becoming the preferred equipment for many for everyday use – family snaps, holidays, sports events – anything. Its common availability belies its potential for many hobbies and for astronomical photography in particular. It doesn't necessarily replace other cameras but rather provides another window of opportunity for amateurs to gather great pictures.

Figure 2.5.

The instrument used for the present work was an Olympus C-300 Zoom (figure 2.5) but many manufacturers supply cameras with similar features. It has 3 million pixels. The number of pixels, or picture elements, determines the resolution or sharpness of the image obtainable. The more pixels the larger the final photograph can be printed without appearing blotchy. Even down to half a million pixels you are still likely to obtain interesting, but small, pictures. A common standard memory, flash or smart-card provided at purchase might well have 16 Mb of storage media. I recommend at least 64 Mb to give great potential for an evening's viewing. A hundred or more pictures can then be taken at high resolution to produce stunning pictures of at least postcard size and often much greater. Of particular use are the following functions.

Selection of Image Quality

The highest quality image is produced through selection of uncompressed TIFF data (your supplier will explain this for your particular camera if you are unsure) around (typically) 1984 × 1488 pixels for a 3 million pixel camera. However, you may only be able to acquire seven images on a 64 Mb card or one on a 16 Mb card. While there may be occasions when this is used, the images in this book were taken selecting some compression to allow many pictures to be taken. You will need to experiment.

Shutter Delay or Self-Timer

Squeezing the shutter release manually causes some camera wobble however hard you try to avoid it. Some pictures that you take will involve much magnification of the image seen by eye and the tiniest movement while pressing the button will be amplified to spoil the picture. Selection of shutter delay will allow a completely hands-off operation of the shutter a few seconds (10 seconds for the Olympus camera described here) following pressing the operate button.

Flash

However bright, almost "blinding", the view appears to you, a camera brightness sensor will often set itself automatically to flash because what it actually "sees" is low overall light. It is therefore important to have the facility to switch it off.

Monitor or Rear Viewing Screen

Images can easily move away from your carefully set target position. Having no way of being aware of the movement can be extremely frustrating. In addition, images just taken can be viewed on-site and immediately to alert you to less acceptable pictures, which can be repeated straight away. It is annoying to find this out later when all your equipment has been put away.

Spot-Metering

This is a highly desirable function that exposes for the brightness within the small frame on the viewing screen to avoid disastrous over- or underexposure. The nice full or partial Moon in the center of your view might seem bright but it is surrounded by darkness and the camera, unless you engage spot-metering, will calculate an "average" brightness to produce an overexposed white blob. There are occasions, however, when it is preferable to have an average exposure to enhance specific features.

Manual Exposure Compensation

This is a feature that is not essential but can assist greatly when it is desirable to increase or decrease brightness through manual selection of set amounts in addition to, or in place of, those set automatically.

Zoom

The zoom function can help in many ways. Zooming in on the view prior to taking a picture can reveal tiny details to confirm you are on target. It can assist with focusing but that method has its drawbacks. Similarly the zoom can allow inspection of a picture already taken to help with deciding its acceptability prior to erasing or loading onto a computer. Zoom will help to expand the image of small objects, such as planets, to fill the automatic exposure selection frame. Of course, this refers to optical zoom as digital zoom has a more limited use.

Panorama

This will allow stitching together of individual pictures downloaded onto the computer (with the appropriate software) but you will need to ensure you have the compatible memory card or the function might not be available.

Batteries

More than nice to have are plenty of rechargeable batteries. One set and some-times two are used for an evening's snapping – very expensive if you are buying non-rechargeables. Decide the best approach for the type of battery(ies) in your camera.

Picture Storage and Retrieval

Within a very short time you will have accumulated thousands of photographs and it is vital to have some way of finding on your computer particular events or individual pictures and details associated with them. As each picture is taken, many cameras automatically record the properties such as shutter and film equiv-alent speed but you must set the clock time yourself initially to be able to retrieve the exact time each photograph was taken – very important for discussions and interpretation possibly many weeks, months or years later. Perhaps store each session's viewing under a separate computer folder labelled with the date and maybe some notes typed into individual picture files. Hours of time could be wasted without at least some order to your system.

Telescope

The objective of this book is to enable everyone to achieve super astronomical pictures of which to be proud. With this in mind the telescope used and recom-mended is a (second-hand is fine and cheaper) 200 mm reflector, although a full discussion of budget telescopes is provided in a book by Patrick Moore and John Watson and referenced in the Appendix. It provides a good balance between economy, performance and transportability. Larger scopes are too heavy to easily carry in and out of the house – or transfer by car to a local observing spot. Without very careful selection, smaller scopes are not as likely to produce amazing pictures although some suggestions are given later.

Figure 2.6.

Described here is the Celestron C8 Schmidt–Cassegrain reflector (figure 2.6) (let's just call it a C8 for now) although many other good makes are available. If you have the money a new one is ideal but with a little patience and effort a second-hand one can be found, especially in the advert pages of astronomy magazines. The C8 used to produce images in this book was second-hand, a discontinued line and about a third of the price of a new one. It was broken in places, but not catastrophically, and some screws were missing. It had no manual but one for a similar model was kindly sent over the Internet by the manufacturer. Also the legs were not telescopic, leading to difficulties in levelling the telescope. None of these adverse factors mattered for the technique described. Particularly advantageous is the celestial tracking mechanism that is part of the C8 and most other instruments of this size. A built-in motor allows the telescope to compensate for the movement of the Earth, thus maintaining the target image within view. For the tasks described here it is sufficient to line up the telescope to face approximately north before commencing a session. More accurate alignment will be required for more advanced astrophotography.

If you can only have one eyepiece for your telescope then a 25 mm will give you most of what you want – especially if your images are downloaded and edited on a computer (see later). Next most important is a ×2 then a ×3 Barlow lens for extra magnification. Discounted lenses can usually be purchased at astronomy exhibitions or negotiated with dealers.

Of course it is great to stand out in the cold clear air acquiring images but also great is taking pictures of the Sun on a nice warm day either waiting for a good show of sunspots or particular events such as occultations (when one astronomical object passes in front of another to hide it from view), transits (when a small object traverses in front of a larger one) and eclipses. Therefore, a great and inexpensive accessory for this purpose is a solar filter made from aluminum-coated glass or plastic. This must be of the type that covers the front lens. It filters out most of the light, letting through just 0.001% of the solar radiation.

<u>Warning!</u>

Do not view the Sun directly without the appropriate filter.

Do not use eyepiece solar filters – they may crack, let through the solar radiation and permanently damage your eyesight.

Do not leave solar viewing telescopes unattended. Casual observers, especially children, will be tempted to view.

Do cover up the finder scope or apply an additional filter.

Do take the advice of an experienced solar astronomer before attempting it yourself unless you are absolutely sure of what you are doing.

Very expensive filters are required to view surface features of the Sun, such as prominences and flares, so these will not be covered here. More on Sun pictures later.

Also useful is a smaller 60–70 mm telescope. The one used for some pictures here is the Meade ETX70AT (figure 2.7).

Filters for the eyepiece are available to bring out the features of a bright Moon and other night sky objects but you can experiment with those as you gain more experience and become more ambitious. None are used here.

Figure 2.7.

CHAPTER THREE

The Magic Ingredient

Overview

Not just magic but very simple. Not just inexpensive but cheap! It is a holder to attach a camera to the telescope while taking pictures. I have attended talks that culminate in the showing of just a few images. Fine for discussing the intricacies, theories and details but not so good for an enjoyable picture show.

One can make many mistakes during an evening's viewing and photographing – with focusing, exposure, telescope wobble or merely forgetting to set or alter a parameter. What is required is the opportunity to take a great many pictures during a single session when many settings can be changed and updated immediately after replaying images from the camera. Statistics also operate well – the more pictures taken the more likely you are to find stunning images amongst them. A hundred or more within a couple of hours is a good session.

There are camera-to-telescope holders available for purchase from astronomical equipment suppliers. However, one in particular that I purchased is dedicated only to one eyepiece at a time. It really must be set up indoors to avoid the structure collapsing and your having to scour the ground in the dark for tiny dropped grub screws. Adjustments during viewing also require considerable skill and practice. The holder described here is homemade and robust. Essentially it consists of three bits of wood and a cheap camera tripod!

To illustrate the ease of build, look at this picture of the construction that facilitated the capture of images produced in this book (figure 3.1).

Telescopes from different manufacturers differ in design. The holder described here is for the Celestron C8 telescope but can be used with others having top accessory screws available. A design to fit those without access to top screws will

Figure 3.1.

be described at a later date, although, clearly, once you understand the general approach you may well be able to concoct your own.

Considerable emphasis is placed on ease of construction. You will see that most components are found amongst odds and ends around the house, garage, workshop or DIY store. In this way the construction can be carried out quickly and cheaply. Consequently, the bits and pieces that you use will differ somewhat from those described and require your creativity to achieve a contraption peculiar to your style of DIY. The assembled structure is attached to, and supported by, the telescope tube.

Parts Required and Method

You will need three metal brackets about 3 cm wide and 2–3 cm each arm length. Slotted holes in one arm are helpful for fine adjustment of final position. Loosen

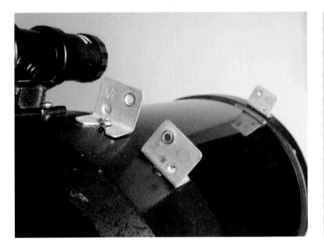

Figure 3.2.

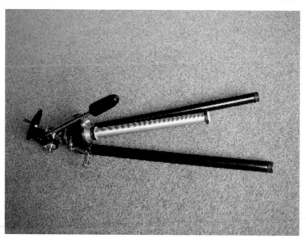

Figure 3.3.

the three telescope top screws and attach a small square of insulating tape around the base to protect the telescope and prevent the brackets slipping during use. Fix the brackets as shown (figure 3.2). Although it doesn't matter which way round they are attached, having one facing the opposite direction to the others counters any slippage during high-elevation viewing.

You will also need a standard camera tripod having sideways, dip and extension control (figure 3.3). It should not carry any surplus weight. For instance, unscrew or otherwise remove leg extensions to leave the upper sections. Only two upper leg sections are required for anchoring, so remove the third one also if possible.

Obtain two pieces of wood (as light as possible but strong) each approximately 60 cm long, 10 cm wide (the width must match the width of the reduced tripod) and 1–2 cm thick. Screw both together, with countersunk screws, to give an overlap of 8 cm (figure 3.4). The exact measurement of overlap will accommodate your tripod's adjustable distance required to position the camera over the eyepiece. Screw three angle brackets, each arm being 2–4 cm in length, into positions

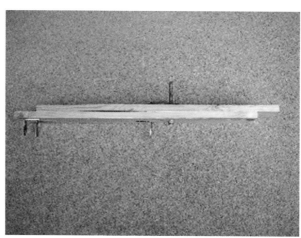

Figure 3.4.

Figure 3.5.

to marry with the telescope top screw brackets (figure 3.5). Drill a hole through the center of the wood about half-way along to take any size bolt you have although it must be long enough to reach above the tripod legs.

Obtain a strip of wood 50 cm long, 4 cm wide and about 0.5 cm thick, which must, once again, be as light as possible but strong. About 2 cm in from one end drill a hole in the center approximately 0.7 cm in diameter. Two holes are shown in the illustration – to accommodate different cameras (figure 3.6). Cut out or drill a slot 20 cm long starting 5 cm from the same end as the drilled hole(s). The slot should be no less than 0.7 cm wide but a little wider will not hurt. You might be able to cut a neater slot than that shown but it was quick to produce. You want to get on with taking stunning pictures as soon as possible, don't you!

The picture (figure 3.7) shows some of the small components required. In particular notice that they are just bits that any DIY hoarder will have collected from broken items (the brackets were salvaged from discarded kitchen cupboard doors).

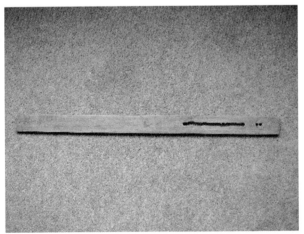

Figure 3.6.

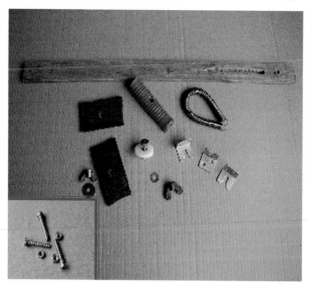

Figure 3.7.

Figure 3.8.

Place the reduced tripod on the wood base so that the shoulder rests on the lower plank of the overlap and nudges hard against the upper plank. Strap the tripod down firmly using a strong spring. Velcro or other material of your imagination can be used but be sure nothing will slip out of place when held in any position (figure 3.8).

Anchor the tripod legs using the bolt and nut – a wing nut is more convenient – through the center of the base. Above and below the legs place a material to cushion the compression and place a load-spreading material on top to avoid damage when tightening the bolt. The example shown comprises two strips of rubber strapping and an appropriately sawn-off discarded plastic handle (figure 3.9).

The structure is now ready to attach to the telescope. Make absolutely sure that connections are all firm and nothing will slip. That would be a disaster during a

Figure 3.9.

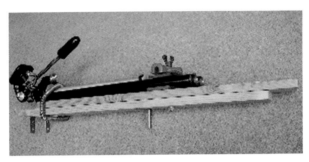

Figure 3.10

photographic session (figure 3.10). Over a soft surface test it by turning it over and upside down quite vigorously.

Attach the prepared thin strip of wood to the face of the tripod camera platform. Replace the standard bolt provided with the tripod by a much longer one and pass it through the slit in the wood. A wing nut is highly preferred as a

Figure 3.11

Figure 3.12.

retainer since constant changes in position during photographing sessions will be required. A small bolt is required of the same thread as that in your camera base. Add a nut (a large plastic one is ideal for ease of use) to act as a lock nut and pass the bolt through the end hole in the wood. This will hold the camera in place (figure 3.11).

Position the structure above the telescope tube and thread appropriately sized screws through each of the three aligned telescope and angle brackets. Tighten the nuts securely (figure 3.12).

The completed structure should look like that shown in Figure 3.13.

The long wooden base is useful for safe parking of the construction. Grab the two top front brackets firmly to take the weight of the telescope, release the declination (dec) clamp (see glossary) and gently allow the telescope tube to tilt forward until the wooden base rests against the wedge.

The contraption that you produce may be heavy or light, depending on the type of wood and tripod used, but do ensure that the declination clamp is firmly tightened during use.

Figure 3.13.

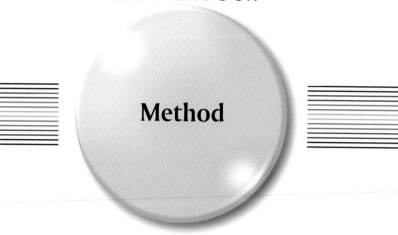

CHAPTER FOUR

Method

Set-up

Having constructed the camera holder (magic ingredient) the camera itself must now be attached. It is strongly recommended that you practise all operations from the comfort of your house or garden on a nice day. Although not at all difficult you must be able to set and move the camera and eyepieces competently in the dark. The following method uses a distant view of a telegraph pole, which also has the advantage of the telescope being able to be placed in a level and stable position and pointing horizontally. Here is the picture using the camera only – no telescope (figure 4.1). Can you see the pole on mid-horizon?

Using the appropriate screw into the base of the digital camera, attach the camera to the strip of wood as shown previously but tilt it upwards until required. Insert a low-power eyepiece into the telescope and focus on the pole. Use spectacles if normally used for distance seeing to avoid blurred photographs. If it is not possible to see through the telescope without bumping your head on the camera, loosen the star diagonal screws (see appendix), turn the eyepiece to point away from the camera and retighten the screws.

Having focused on the pole turn the star diagonal back to position the eyepiece in line with the camera. The view will remain in focus. Be very sure that the screws of the star diagonal are firmly tightened. Give it a slight wobble whilst holding to make sure it doesn't drop and further tighten if necessary. It is wise to place a cushion or other soft surface below the star diagonal. Without warning it can fall to the ground if it has not been sufficiently tightened.

Open the camera (turn it on) so that the lens protrudes as for normal operation and use the tripod tilt control to bring the camera over the eyepiece. If it is too

Figure 4.1.

close or far then use the vertical adjustment (wing nut) to slide the camera up or down the slot in the strip of wood – one hand on the camera, the other on the wing nut. The camera lens must hover as close as possible to the eyepiece. Turn on the camera back screen view and adjust the tripod forward–backward and sideways swing to center the telegraph pole image on the screen. It is vital that this operation is practised to become adept.

Set the camera menu to disable the flash (of course) and add the shutter delay. Take a picture (figure 4.2). Do not move during the shutter delay period! Even walking on the concrete or carpet (wherever you are) will create some vibration. Watch the camera screen. For some cameras for about a second following shutter operation your actual picture captured will appear for your assessment. This avoids having to close or turn off the camera to view and start the settings all over again for the next picture and is very useful during a night sky session. Variations of this facility are available on most cameras for immediate viewing. If you are not satisfied with the image make any necessary adjustments and take another, and another …. You have got a digital camera after all that will take many pictures. Here the pole picture was taken using a 40 mm eyepiece. (See the Appendix for an explanation of strengths or magnification power of eyepieces.)

It is important to understand the orientation of pictures captured. A reflecting telescope reverses the image left to right, that is, the object appears the right way up but if you could see someone walking to the right they would appear to be walking to the left. Now visualize where you would have to stand (or hover!),

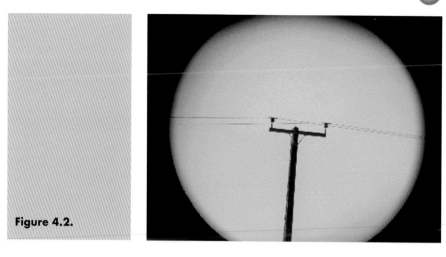

Figure 4.2.

when your camera is positioned over the telescope to take pictures, to look through the camera sighting window (or on screen). Left and right would now be normal but the image would be upside down! The net effect is that all you need to do on the computer to produce a normally orientated picture is to flip through a horizontal axis. This has been done for the pictures reproduced here. Ensure you fully understand this by observing your own pictures and comparing them with the direct telescopic and binocular view.

Tilt the camera away from the eyepiece, turn the star diagonal away from the camera and choose a higher magnification eyepiece, say 25 mm. It will be necessary to raise or lower the camera along its holder to allow for the varying length of the lenses and must be practised ready for use in action. Take another picture (figure 4.3).

Now add a ×2 Barlow to your 25 mm lens and repeat the set up as above (figure 4.4). It is quite remarkable to see the magnification of a land object through an astronomical telescope. It is easy to forget the power when viewing the sky. Here

Figure 4.3.

Figure 4.4.

Figure 4.5

is a succession of images taken through higher strength lenses up to 10 mm plus a ×2 Barlow (figures 4.5 to 4.7).

Compare the highest magnification image with the original view without telescope where the pole could hardly be seen. Of course, atmospheric distortions (boiling effect) interfere grossly with terrestrial images at high magnification.

Another word of warning. Whilst practising the above technique be very careful not to swing the telescope towards the Sun without the appropriate solar filter.

Many cameras having automatic exposure will adjust for the average brightness over the whole screen. This is very difficult to work with though not impossible. Spot-metering is highly desirable, particularly for the Moon, Sun and planets, so practise using this also.

Select the resolution (quality) of your pictures. The higher the resolution the better the image but the fewer stored in the camera for a single session. For a 64 Mb memory card the highest resolution will only allow seven pictures while

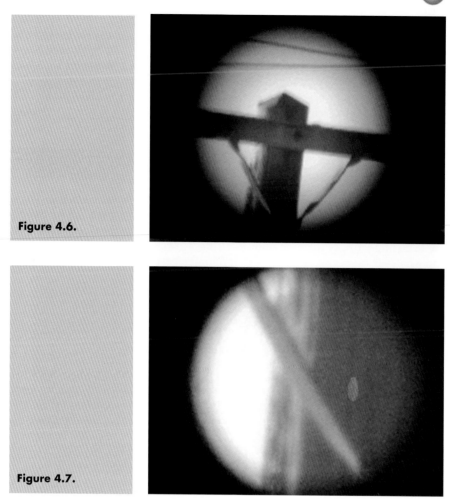

Figure 4.6.

Figure 4.7.

the lowest accumulates over 600. Photos in this book were taken using HQ 1984 × 1488 giving a capacity of approximately 80–100 pictures. You will need to experiment to find the most appropriate resolution for you and your camera.

Camera technology advances rapidly. A 3 megapixel (Mp) camera used to capture images here allowed great pictures with little noticeable difference in resolution between higher pixel cameras – up to around A4 size when printed. However, 8 and even 12 Mp are now available and will allow much larger prints but come at a greater cost.

Many camera features are automatic although available for manual setting. For film cameras a film is selected for its speed, i.e. ISO 100, 400, 2,400, etc. The digital camera allows simulated selection of ISO number either manually (100, 200,400 for the Olympus) or through a much finer choice of automatic settings. So, for example, an ISO of 60 might be automatically set. While becoming familiar with the complete operation, automatic settings are to be preferred especially since the camera combines selection of the best combination of shutter speed, aperture and ISO. Manual ISO settings of 100, 200 and 400 were found to provide inferior images to that set automatically for both picture resolution and "graininess".

It is worth repeating that rechargeable batteries are a real advantage since one session of around 100 snaps plus downloading and erasing will drain a set. Therefore several sets are required with constant recharging to allow pictures to be taken every night and/or day.

Focus

The focus is by far the most important parameter in obtaining the best pictures possible. You might wait many days, weeks or months for good seeing conditions, the clarity and stability of the air and lack of high wispy cloud, etc., then waste an opportunity through poor focusing. However many pictures you take and whatever the region of the Moon or Sun, the best pictures will be those with the sharpest focus. Here are a few methods.

Direct

As you view the target it will usually appear as though it is wobbling or boiling and focusing seems to be impossible. Persist to get the best image. Take a picture. Do not forget to wear your spectacles for final focus should you normally require them for long distance or the picture will not be sharp. Go backwards and forwards through the best focus time and again until you have thoroughly noted the clearest details of the target feature. Then adjust the knob to achieve that clearest image.

Bracketing

Often, the optimum focus will be just a little away from what you have set. As the focusing knob is turned small notches can be felt. Make an attempt at best focus. Turn the knob one way to create a noticeable amount of out-of-focus. Turn back through the best focus to achieve out-of-focus the other way and count the number of notches travelled through. Turn the knob back again only this time take a picture at each notch position. One of the pictures will be at the best focus. It might seem wasteful at first glance but that is precisely the advantage of the digital camera. All or any of your unwanted pictures can be unceremoniously dumped.

Screen Focus

An approximate focus can be achieved by viewing the target on the camera screen. However, it is sometimes tricky to decide exactly at which point the focus is best, due to the low resolution of the screen image.

If you are trying to capture a particular feature on the Moon or Sun ensure that feature is in the center of your view to avoid lens edge effects spoiling the main focus. Refocus frequently during the session as imperceptible slippage of some adjustments can occur and atmospheric conditions can change. Also you might not have achieved the absolute best focus first time. Experiment with the zoom facility although the author has not found that to improve sharpness of focus of the Sun or Moon images at all – in fact it often degrades them.

Other Factors

General

To achieve best pictures it is commonly recommended to attend to temperature equilibration by placing the telescope at the point of use for up to 3 hours prior to operation. In addition, stand the scope on grass or soil rather than concrete or stone since the latter will have accumulated much heat during the day that will cause unstable air due to convection currents. Keep as far away from light pollution as possible. But! All of the images presented in this book took no notice of any of these factors. The telescope, stored with its camera cradle already attached, was lifted on to a concrete slab patio, with a glaring street light only 7 meters away, and operated immediately. The first pictures were captured usually within 5–10 minutes, giving enthusiastic amateurs the encouragement to take pictures with the absolute minimum of preparation. All too often lengthy preparations can be most discouraging. Once great images are obtained more advanced preparation and care can be attempted to assess improvement possibilities.

Computer

With a camera full of pictures the next stage is to download them onto a computer. The camera will probably have been purchased with the necessary software included to transfer the pictures to a computer and perform simple editing. The most important and frequently used functions are crop, zoom (not to be confused with camera zoom while taking pictures) and flip. An image caught at good seeing conditions and accurately focused is capable of much computer zooming-in or cropping to present a particular feature or merely to eliminate less acceptable areas of an image. Remember that since the camera is effectively "upside down" the image must be flipped to have the picture in normal or binocular view.

Color

In reality the Moon is a very monochrome world. Almost everything is a shade of gray. But many of the pictures taken also exhibit shades of yellow, brown or even red! Clearly the camera photodetectors will have, to some degree, a different color response to the human eye but nothing too significant. As light passes through

Figure 4.8.

the atmosphere it will be greatly affected (refracted) by the amount of air (the angle or path from object to viewer) and the quantity and type of pollution. For example brilliantly colored sunsets are often witnessed following volcanic eruptions.

The intention of the described approach to photographing sky objects is to capture images just as they are received, including what might be described as color distortion. Look at this first picture that shows a very yellow–brown image (figure 4.8). And a second showing an even more red–yellow hue (figure 4.9).

Figure 4.9.

Figure 4.10.

In fact the air can be very unstable and change its light refraction or absorption on very short time-scales. Often seen is a sharp change from gray to yellow during a night's viewing. One sequence was actually seen to change such colors many times successively for individual shots. Interference from stray light can be minimized to some extent through the use of a dew shield, a shroud to prevent dew forming on the telescope lens. This can be easily made or cheaply purchased.

Fortunately the computer can come to the rescue to allow for individual preferences. At the press of a button, gray-scale can be selected to present the Moon in a more realistic light as shown here (figure 4.10).

It is worth mentioning at this point that there are very pale, almost invisible colors associated with different areas of the Moon's surface corresponding to different chemical elements. One attempt has been made to excessively enhance those colors to produce a beautiful picture of the moonscape.

As always, experiment with factors such as color and brightness to achieve the most acceptable photograph for you.

CHAPTER FIVE

The Universe and You

When deciding on your photographic targets consider this.

How is it that there are so many beautiful sights to see in the night sky? Where did they come from? And what were you doing 14 billion years ago? At that time there was an almighty bang, or so the theory goes, and the beginnings of the Universe were created out of nothing.

Understanding of the Universe has greatly increased in recent years. The current model of how the Universe formed is known as the Big Bang (figure 5.1). Before that the steady state theory of cosmology claimed that the Universe simply existed without changing with time, a theory that presents many difficulties. Evidence suggests that the Universe is expanding. While there are ways to explain expansion in

Figure 5.1.

a steady state Universe there is little evidence to support it. The big bang theory states that at some time in the distant past there was nothing. A process known as vacuum fluctuation created what is known as a singularity. From that singularity, which was about the size of a large pea, our Universe was born.

It is hard to imagine the very beginning of the Universe. Physical laws as we know them did not exist due to the presence of incredibly large amounts of energy, in the form of photons. Some of the photons became quarks, and then the quarks formed neutrons and protons. Eventually huge numbers of hydrogen, helium and lithium nuclei formed. The process of forming all these nuclei is called big bang nucleosynthesis. Theoretical predictions about the amounts and types of elements formed during the big bang have been made and seem to agree with observation. Furthermore, the cosmic microwave background (CMB), a theoretical prediction about photons left over from the big bang, was discovered in the 1960s and mapped out in the early 1990s.

After some period of time following the big bang, gravity condensed clumps of matter together. The clumps were gravitationally pulled towards other clumps and eventually formed galaxies. It is difficult to model how this clumping may have occurred, but most agree that it occurred faster than it should have. A suggested explanation is that right after the big bang the Universe began a period of exaggerated outward expansion, with particles flying outward faster than the current speed of light, an explanation known as inflation theory that has widespread support.

Galaxies then formed containing unimaginable numbers of stars and star clusters (figure 5.2). And since they formed from matter that was moving rapidly, they also moved rapidly. Due to a phenomenon called Doppler shifting, the wavelength emitted by something moving away from us is shifted to a lower frequency, and the wavelength of something moving towards us is shifted to a higher frequency just like a police siren sounding higher as it approaches us and lower as it goes away. The result is that visible wavelengths emitted by objects

Figure 5.2.

moving away from us are shifted towards the red part of the visible spectrum, or redshifted. And the faster they move away from us, the more they are redshifted. Thus redshift is a reasonable way to measure the speed of an object. When we observe the redshift of galaxies outside our local group, every galaxy appears to be moving away from us. We are therefore led to the conclusion that our Universe is expanding. This is called Hubble expansion, after Edwin Hubble, who discovered the phenomenon in 1929.

The Oscillatory Universe model claims that the Universe started with a big bang, and that it is currently expanding. Eventually, however, the expansion will slow, stop, and then the Universe will begin to contract. The contraction will continue until all of the mass of the Universe is contained in a singularity and referred to as the Big Crunch. The singularity then undergoes a big bang, and the process begins afresh. It must be emphasised that the Big Bang is just current theory and not all cosmologists are comfortable with it. So we can expect at least some refinements and modifications if not a completely new theory at some time in the future. The question of whether the Universe will collapse in a big crunch or carry on expanding forever continues to be posed but where do we fit into all of this?

It is believed that the Solar System was formed when a cloud of gas and dust in space was disturbed, maybe by the explosion of a nearby star, a supernova. This explosion made waves in space that squeezed the cloud of gas and dust. Squeezing made the cloud start to collapse, as gravity pulled the gas and dust together, forming a solar nebula. The cloud began to spin faster as it collapsed. Eventually, the cloud grew hotter and denser in the center, with a disk of gas and dust surrounding it that was hot in the center but cool at the edges. As the disk got thinner and thinner, particles began to stick together and form clumps. Some clumps got bigger, as particles and small clumps stuck to them, eventually forming planets or moons. Near the center of the cloud, where planets like Earth formed, only rocky material could stand the great heat. Icy matter settled in the outer regions of the disk along with rocky material to form the outer planets and between them material attracted and held onto the hydrogen and helium gas of

Figure 5.3.

the solar nebula to form the Jovian planets (figure 5.3). As the cloud continued to fall in, the center eventually became so hot that it became a star, the Sun. After 10–100 million years or so the proto-Sun ignited violently and for the next tens of millions of years was very active and blew away the left-over gas and small solid particles from the solar nebula, terminating the planet formation process. By studying meteorites that were thought to be left over from this early phase of the Solar System, estimates of its age have been put at about 4,600 million years.

So now we know where we are and where we came from.

It can be difficult to visualize the arrangement of planets and the Sun. Should you be passing the village of Otford near Sevenoaks in Kent, help is at hand.

A millennium project devised by David Thomas was constructed to represent the planets and their position in relation to the Sun as they stood on 1 January 2000. It is the largest scale model of the Solar System in the world and beautifully laid out in the country environment centering on the local playing field. Robust white pillars support stainless steel disks on which are represented the planets and their size relative to the central shiny hemispherical Sun. To achieve the necessary scale, the playing field soon became too small and Saturn onwards are represented around the village with Pluto amongst the fields on the horizon. David Thomas is exceedingly enthusiastic in its construction and future. The nearest star is 4.2 light years away and represented in Los Angeles! Not only are similar locations on Earth being considered for other stars but the Moon and …? Wow! Here, (figure 5.4), David uses a homemade planet-position calculator placed over the Sun.

Figure 5.4.

Figure 5.5.

Earth can be seen just a short walk from the Sun (figure 5.5). Mars is a plate flush with the playing ground (figure 5.6), rather than a pillar as for all the others, including the indicator of orbit and direction mown into the grass. Also shown is

Figure 5.6.

a close-up of Jupiter (figure 5.8), a view towards the hills where Pluto stands (figure 5.9) and a slightly wider snap of Venus plus Earth and Mercury (figure 5.7).

Lots more detail and leaflets are available from the local Heritage Center.

At least one other large-scale model of the Solar System is being planned in the UK.

Figure 5.7.

Figure 5.8.

Figure 5.9.

CHAPTER SIX

Targets

Overview

Different objects in the sky often require very different techniques. To consider photographing them all could drown an amateur, new to astrophotography, in detail, expense and confusion: nebulae, galaxies, the Moon, the Sun, planets and stars. I consider the most amazing sight to begin with is the Moon. It is also regularly accessible (weather permitting as usual) and has a plethora of fascinating features. Also, with the same equipment plus filter, the Sun is interesting to monitor and to be ready for the unexpected and sometimes spectacular sunspots. These two targets alone would be sufficient to achieve beautiful results. However, with a little more patience and attention to detail very satisfying pictures of some planets can be obtained, and is described.

The Moon

To get the first pictures – wait for a clear night with the Moon at any phase, select a low-magnification eyepiece, carefully aim and shoot. Replay the image onto the camera screen and see what a great picture you have taken! Shoot a few more, possibly using different eyepieces, and load images onto your computer. The first sight of the amazing detail and features of the Moon are stunning. A whole new world appears (figure 6.1). Wow!

The appearance and position of the Moon is often taken for granted but in fact the path of the Moon relative to the Earth is quite complicated and only retraces

Figure 6.1.

its exact path after approximately 18 years. Therefore the Moon sometimes appears low in the sky and sometimes high. When low, the light must pass through more of the Earth's atmosphere causing more distortion and poorer quality pictures. Its orbit is at 5 degrees to that of the Earth's. It also has been gravitationally "captured" by the Earth and therefore always presents the same side to Earth. Since it is spherical (almost) the most that can be seen of the surface at any one time is 50%. However, it also wobbles such that photographing at different times covers a total of 59% of the lunar surface. The "wobble" is called libration and there is a longitudinal as well as a latitudinal libration. Other smaller effects occur to expose slightly more of the Moon's face towards Earth and your challenge may well be to capture features at the edge of the Moon to cover as much of the 59% as possible.

The Sun

Never, ever look through the telescope without the correct solar filter in place! One momentary slip and you will certainly do huge damage to your eye – probably resulting in blindness in that eye. Look at this picture (figure 6.2).

At the close of a Sun-observing session (figure 6.3), the eyepiece was removed and the protective cap replaced. The solar filter was removed and the front protective shield replaced, which took about 5 seconds or less. But! The scope was still pointing at the Sun! During those few seconds the thick plastic lens cap had burnt through and burst into flame! Imagine the effect on your eye! Or your eyepiece! Or your camera!

Always leave your finderscope cap in place (on the front of the scope to avoid damaging the finderscope as well as the observer). It will similarly concentrate

Figure 6.2.

Figure 6.3.

the Sun's rays and quickly burn your ear, shoulder, clothing or anything else in line. Never leave a telescope unattended. Any curious bystander might have a "quick look" especially a child. Also refer back to Sun filters in the Equipment chapter 2.

Having addressed all the safety precautions point and shoot as for any other subject, the weather being the usual annoying variable.

The Planets

In this amazing age of space awareness we think of planets as being in our own backyard. Images sent back from space telescopes and probes show minute detail of their atmosphere and surface (figure 6.4). And yet our nearest planet, Venus, in our Solar System is still 50 million kilometers away. And you can capture pictures of it using the techniques described in the following pages. You can even

Figure 6.4.

capture those of Saturn with satisfying detail of the rings and more. And that is around 1,300 million kilometers away! Acquiring such images needs a little more attention to detail and practice than for the Moon but is not difficult. The rings of Saturn and the cloud bands and moons of Jupiter really are there. Venus really does have phases as does the Moon. You can see them and record their existence.

Other Objects

With practice other sky objects become available to your camera but it is recommended that these are challenges attempted as expertise increases following the production of hundreds or thousands of sharp Moon, Sun and planet images. The exception to this approach is the collection of pictures accumulated, not for their staggering beauty but to record your observation of particular events or sky objects – to prove that you witnessed them albeit with just discernable images. Comets and nebulae come under this heading. Of course, with suitable software applied to multiple images even these can be racked up to wonderful pictures.

CHAPTER SEVEN

Our Moon

Where Did It Come From?

Mankind may have stared in wonder at it, worshipped it and studied it but at some point in time someone started the ball rolling. Where did the Moon come from and how was it formed? Did it arrive from somewhere? Has it always been there? Will it always be there? Without knowledge, for thousands of years, the Moon's origin could only be guessed at and incorporated into the prevailing mythological or religious beliefs (figure 7.1). The question could only be posed, not answered, until the birth of astrophysics as we now know it. Modern scientific study of our satellite really started in 1610 with Galileo pointing his telescope towards the Moon and describing the dark and light regions as vast plains and rugged mountains. George Darwin, son of the famous Charles Darwin, proposed the first credible theory in 1878 of where the Moon came from. A large chunk of the Earth was ripped away due to the rapidly spinning Earth. Soon after, a geologist, Osmond Fisher, stated that the Pacific Ocean was the origin of the chunk.

If we exclude some very unlikely ideas such as multiple collisions at the same or different times in the evolution of the Earth, hypotheses essentially crystallize into the following possibilities:

1. A part of the Earth was somehow ripped away, the suggested spot for this being the Pacific Ocean Basin. It was possibly hurled into space from a rapidly spinning Earth.
2. The Moon might have been born elsewhere in space and gravitationally captured as it passed close to Earth. Proposed in 1909 by Thomas Jefferson Jackson See.

Figure 7.1.

3. Both Moon and Earth formed together from the original nebula, or gas cloud, that formed the Solar System. Proposed by astronomer Edouard Roche, and others.
4. Planetesimals on unstable trajectories that orbited the Earth and Sun collided to form clouds of smaller particles and dust that condensed to form the Moon. It was then captured by the Earth.
5. A large wandering body the size of Mars flew into the early Earth soon after the Earth's formation. It delivered a glancing and spectacular blow. The resultant scattered material coalesced or recombined to leave the Earth with a gained satellite. Proposed by Hartman and Davis in the mid-1970s.

Since retrieval of samples from the Moon's surface by manned visits to the Moon, much in the way of facts and evidence has become available. The composition of Earth and Moon samples has been compared. Because the materials of each have major differences, and it doesn't have a significant iron core as does Earth, it has reduced the possibility that both were formed from the same substrate at the same time. On the other hand, having major similarities makes it unlikely that Earth captured a wandering visitor from elsewhere in the Solar System or space beyond.

The most favored theory is currently the impact of a large Mars-sized chunk of rock or planetesimal from elsewhere in space and is often referred to as the "Big Whack". At the time Earth formed 4.6 billion years ago, other smaller planetary

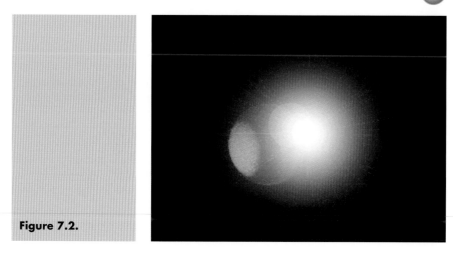

Figure 7.2.

bodies were also growing. One of these hit the Earth late in the growth process, blowing out rocky debris into orbit around the Earth and aggregated into the Moon. The theory emerged in 1984 following the analysis of rock and "soil" samples brought back by astronauts between 1969 and 1972. 382 kg of rocks and regolith (the layer of dust and bits of rock covering the surface of the Moon) have been brought back by nine space missions. Six Apollo manned landings (see Chapter 11 for pictures of landing sites) and three Soviet Luna robotic missions accomplished the collection (figure 7.2). The big whack suggested that around 50–100 million years after the Solar System was formed a huge protoplanet crashed into the still molten Earth, heating both objects and throwing out colossal amounts of debris. A portion of the resulting thick ring of vapor, dust and lumps of rock came together or accreted to form larger pebbles, rocks and planetesimals that collided and fused to form the Moon. None of the older theories explained how the Moon's oldest rocks solidified from molten rock about 4.44 billion years ago, roughly 100 million years after the Solar System formed.

If the Moon had been captured ready-formed from elsewhere it would most likely have entered a highly elliptical orbit. The ocean tides would then have been very widely spaced and sand and mud deposits would have shown long intervals between them. However, the oldest record of Earth's ocean tides comes from South African rocks over 3 billion years old in which were revealed very evenly spaced sand and silt layers deposited by daily, fortnightly and monthly tidal cycles – a consequence of a more circular orbit. A highly elliptical orbit would have caused the rocks to have some sand layers from the few days each month when the Moon was close enough to cause tides, then a thick layer of mud representing the lack of tides when the Moon was far from Earth. It is interesting also that at that time the Moon orbited Earth much faster and was significantly closer.

Studies of radioactive decay of niobium 92 to zirconium 92 and hafnium 182 to tungsten 182 imply a big collision formed the Moon at least 50 million years after the birth of the Solar System. Only later did Earth and the Moon develop all their distinct layers. Oxygen isotope ratios are identical for the two bodies, therefore making the capture theory less likely than a time of common formation.

The impact idea explains why the Earth, with its iron core, is heavier for its size than the Moon, the density of which more closely resembles that of the Earth's mantle rather than the Earth as a whole. Had they formed at the same time the Moon would also have a significant iron core. The heavy iron core of the impacting planetesimal would get trapped and incorporated into the Earth while only the lighter rocky materials would get blown out into orbit creating the Moon. It also accommodates the known facts concerning the angular momentum of our system – the impactor's smack had to be of a size and angle that would produce present-day spin. An object larger or smaller than Mars would not account for the composition and spin of the Earth and Moon. It even accounts for our planet's odd, 23.5 degree tilt off the ecliptic plain. The rarity of volatiles (for example, water, nitrogen and carbon dioxide) on the Moon is also explained, unlike the other theories.

Computer models have been attempted to incorporate factors of melting, vaporization, gravity and so on. The largest step forward was the arrival of an approach called Smoothed Particle Hydrodynamics (SPH), which begins by treating the early Earth and the impactor as many small particles. The evolution of the particles under Nature's forces is then followed. Only the availability of powerful computers allows the input of a sufficient number of particles (many thousands) to produce meaningful results. Jay Melosh of the University of Arizona pointed to the complexity of the calculations to avoid jumping to conclusions too soon: "Not only does such a collision involve all the details of shock physics, melting and vaporization but the mutual interaction of all those hot fluids squirting around in space have still to be taken into account." How marvellous that man has been to the Moon to collect samples with which to compare with Earth and create more informed theories. Even more marvellous that we are used to astronomical bodies taking millions of years to move or form, but the Moon is stated to have been formed within a year!

Finally we must say that everything is not completely tied up. So maybe it was a little premature to have actually given the impactor a name – Theia! There are still tantalizing questions and we may yet get surprises as more data are collected. The European Space Agency's ion propulsion lunar probe SMART-1 has begun returning stunning images such as those of craters at the limb, Brianchon, Pascal, and Pythagoras. Calcium and other elements have also been positively identified.

The Surface of the Moon

Galileo, with his first telescope, saw that the Moon was not the perfect ball assumed by many. He noted, through his less than perfect lenses, dark and light areas and some large craters. A modern image of the Moon, figure 7.3, clearly illustrates these expanses of dark and light areas. He published five of his drawings of the whole disk in a small volume that he called Sidereus Nuncius or The Starry Messenger. He wrote "I saw the Moon from such close proximity as if it were barely two terrestrial diameters distant. The surface of the Moon is neither smooth nor uniform. Nor very accurately spherical. It is uneven, rough, replete

Figure 7.3.

with cavities and packed with protruding eminences, in no other wise than the Earth which is also characterized by mountains and valleys."

Following the formation of the Moon as a distinct and separate entity, its outer layers were molten due to the heat produced from the collision. The melt was probably very deep, possibly several kilometers. For about 500 million years it, and other planets and moons, was bombarded with lumps of rock still flying around at that time. Large projectiles hitting the Moon created the major basins of Mare Tranquillitatis and Fecunditatis and soon after, the expansive Mare Imbrium. As the Moon cooled a crust formed on the outer layer but the bombardment continued. Large pieces of debris impacting on the surface at high speed carved out huge craters that then filled with molten magma released from just below the surface. Plato and Grimaldi are more flooded basins than impact craters. As the Moon cooled and lava flows ended the continuing bombardment carved out craters large and small. These craters remained intact, unflooded and preserved until the present day. Erosion was absent due to the lack of forces caused by an atmosphere or crust movement as occurred on Earth. The younger craters such as Tycho and Copernicus therefore retained their detail – the high mountain edges, central peaks and layered walls. Some volcanic activity contributed to features such as rills or cracks in the surface. These and other features such as valleys, domes and rays will be covered later.

The Moon of today is a dead world. The hugely cratered and marked surface sits like a fossilized monument in the sky. Approximately 30,000 craters having a diameter of greater than 1 km cover the near side (figure 7.4). There is virtually no atmosphere, no life, no additional craters since detailed pictures were first captured, and very little movement anywhere on the crust. Undoubtedly there is always the chance of a piece of space debris arriving to form another crater,

Figure 7.4.

which would probably obliterate an existing one to make room amongst all the others. An in-depth description of the composition of the Moon is beyond the scope of this book but there are many excellent works written on the subject some of which are listed in the appendix.

The Moon in Our Sky

It looks very bright in our sky with the naked eye and it is almost blinding through a telescope. Of course, it has no source of light of its own and depends entirely on light from our Sun. Bright as it appears it is a very poor reflector, reflecting only about 7% of the incident light. Reflectivity of an astronomical body is known as its albedo. However, the full Moon reflects enough light for us to see at night although the eye is not sufficiently sensitive to see in any other color than gray with the possible exception of red.

Who has not said "once in a blue moon" to denote an unusual occasion? Can the Moon ever be seen to be blue? Although there are rare times when this might have happened due to atmospheric pollution from natural phenomena the phrase more commonly applies to the occurrence of a second Full Moon within a calendar month.

But where in the sky is the Moon? It seems fairly reliable in appearing where we expect it to appear but never quite in the same place! We say it goes round the Earth and that has been accepted for a very long time. But how does it go round – what path does it follow? In fact it doesn't quite go round the Earth. Since the Moon is very large, it and the Earth actually rotate around a point somewhere between the centers of each body. The point is known as the barycenter, which is 1707 km below the Earth's surface, so for most practical purposes the Moon can be considered to go round the Earth. Although it was stated earlier that the orbit of the Moon round the Earth is circular (that lends support to the big whack theory) it is not quite so. The Moon, in fact, traces a somewhat elliptical path and is said to have an orbital eccentricity of 0.0549. The closest approach to the Earth's surface (perigee) is 348,294 km and the furthest (apogee) 398,581 km. Clearly the Moon will look slightly larger in the sky at perigee than apogee. Many factors determine the route followed by the Moon and it therefore only retraces its exact path approximately every 18 years.

The Moon moves eastwards in its elliptical orbit at the rate of about one diameter per hour, or 13 degrees per day, against the celestial background, accounting for the fact that it rises later each night.

I doubt there are many, even the most casual of observers, who do not recognize that the patterns seen on the Moon appear to be always the same – the same side constantly facing us. The side we do not see is often referred to as the dark side, but, of course, that is only a convenient statement. The Sun illuminates "both sides" equally as the Moon journeys round the Earth. Early in the Earth–Moon history the Moon did indeed reveal both sides as it revolved rapidly. However, the Moon was not symmetrical and the slight bulge towards the Earth received a stronger pull from the Earth, which acted as a brake, finally capturing the same face towards us. In addition it is a consequence of tidal coupling or locking, an interaction between distortions of each caused by gravity as they rotate about one another. The orbital period, the time it takes to go round the Earth, became the present-day 27.3 days, exactly the same as its rotation period. Of course, the Moon continues to slow down in its orbit due to the tidal locking and will eventually remain visually static over one point of the Earth.

The Terminator and Phases of the Moon

Although most mention has been of the circular shape of the Moon it clearly does not always appear so in the night sky. Like the Earth, the Moon is a sphere which is always half-illuminated by the Sun, but as the Moon orbits the Earth we get to see more or less of the illuminated half. During each lunar orbit (a lunar month), we see the Moon's appearance change from not visibly illuminated through partially illuminated to fully illuminated, then back through partially illuminated to not illuminated again. Although this cycle is a continuous process, there are eight distinct, traditionally recognized stages, called phases. The phases designate both the degree to which the Moon is illuminated and the geometric appearance of the illuminated part. These phases of the Moon, in the sequence of their occurrence (starting from New Moon), are listed below.

New Moon

The Moon's unilluminated side is facing the Earth (figure 7.5). The Moon is not visible (except during a solar eclipse). However, traditionally the first sighting of a thin Moon is also commonly referred to as the New Moon.

Figure 7.5.

Waxing Crescent

The Moon appears to be partly, but less than one-half, illuminated by direct sunlight (figure 7.6). The fraction of the Moon's disk that is illuminated is increasing.

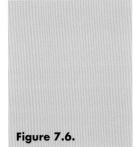

Figure 7.6.

First Quarter

One-half of the Moon appears to be illuminated by direct sunlight (figure 7.7). The fraction of the Moon's disk that is illuminated is increasing.

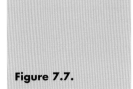

Figure 7.7.

Waxing Gibbous

The Moon appears to be more than one-half but not fully illuminated by direct sunlight (figure 7.8). The fraction of the Moon's disk that is illuminated is increasing.

Figure 7.8.

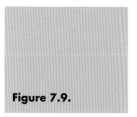

Figure 7.9.

Full Moon

The Moon's illuminated side is facing the Earth (figure 7.9). The Moon appears to be completely illuminated by direct sunlight.

Figure 7.10.

Waning Gibbous

The Moon appears to be more than one-half but not fully illuminated by direct sunlight (figure 7.10). The fraction of the Moon's disk that is illuminated is decreasing.

Figure 7.11.

Last Quarter

One-half of the Moon appears to be illuminated by direct sunlight (figure 7.11). The fraction of the Moon's disk that is illuminated is decreasing.

Figure 7.12.

Waning Crescent

The Moon appears to be partly but less than one-half illuminated by direct sunlight (7.12). The fraction of the Moon's disk that is illuminated is decreasing.

Following the waning crescent is New Moon, beginning a repetition of the complete phase cycle of 29.5 days average duration. The time in days counted from the time of New Moon is called the Moon's "age". Each complete cycle of phases is called a "lunation". Because the cycle of the phases is shorter than most calendar months, the phase of the Moon at the very beginning of the month usually repeats at the very end of the month. When there are two Full Moons in a month (which occurs, on average, every 2.7 years), the second one is called a "Blue Moon".

Although Full Moon occurs each month at a specific date and time, the Moon's disk may appear to be full for several nights in a row if it is clear. This is because the percentage of the Moon's disk that appears illuminated changes very slowly around the time of Full Moon (also around New Moon, but the Moon is not visible at all then). The Moon may appear 100% illuminated only on the night closest to the time of exact Full Moon, but on the night before and night after will appear 97–99% illuminated; most would not notice the difference. Even two days from Full Moon the Moon's disk is 93–97% illuminated.

The phases of the Moon are related to (actually, caused by) the relative positions of the Moon and Sun in the sky. For example, New Moon occurs when the Sun and Moon are quite close together in the sky. Full Moon occurs when the Sun and Moon are at nearly opposite positions in the sky – which is why a Full Moon rises about the time of sunset, and sets about the time of sunrise, for most places on Earth. First and Last Quarters occur when the Sun and Moon are about 90 degrees apart in the sky. In fact, the two "half-Moon" phases are called First Quarter and

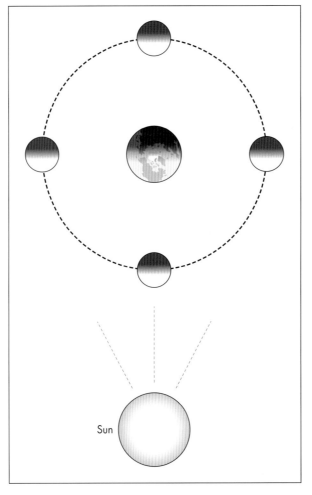

Figure 7.13.

Last Quarter because they occur when the Moon is, respectively, one- and three-quarters of the way around the sky (i.e. along its orbit) from New Moon.

The relationship of the Moon's phase to its angular distance in the sky from the Sun allows us to establish very exact definitions of when the primary phases occur, independent of how they appear. Technically, the phases New Moon, First Quarter, Full Moon, and Last Quarter are defined to occur when the excess of the apparent ecliptic (celestial) longitude of the Moon over that of the Sun is 0, 90, 180, and 270 degrees, respectively. These definitions are used when the dates and times of the phases are computed for almanacs, calendars, etc. Because the difference between the ecliptic longitudes of the Moon and Sun is a monotonically and rapidly increasing quantity, the dates and times of the phases of the Moon computed this way are instantaneous and well defined. The "percentage of the Moon's surface illuminated" is a more refined, quantitative description of the Moon's appearance than is the phase. Considering the Moon as a circular disk, the ratio of the area illuminated by direct sunlight to its total area is the fraction of the Moon's surface illuminated; multiplied by 100, it is the percentage illuminated. At New Moon the percentage illuminated is 0; at First and Last Quarters it is 50%; and at Full Moon it is 100%. During the crescent phases the percentage illuminated is between 0 and 50% and during gibbous phases it is between 50% and 100%. For practical purposes, the phases of the Moon and the percentage of the Moon illuminated are independent of the location on the Earth from where the Moon is observed. That is, all the phases occur at the same time regardless of the observer's position.

From the diagram it is easy to see that a shadow progresses across the face of the Moon as seen from the Earth on successive nights (figure 7.13). The line between the dark and light part is called the terminator and is the most important feature as far as observers are concerned. The Full Moon is marvellous in its brightness and overall features but details of craters, maria and mountains can only be fully appreciated when illuminated at an angle by the Sun i.e. when seen close to the terminator. The image shown in figure 7.14 exemplifies this grandly. First glimpse through a telescope of the rugged and beautiful terrain is a wondrous moment, and to be able to capture such images quickly and inexpensively as described here is a marvel indeed.

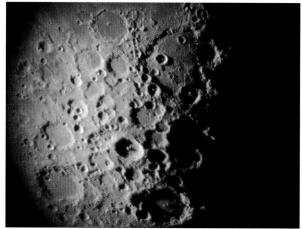

Figure 7.14.

CHAPTER EIGHT

The Moon –
First Glance

Look at the Moon. What do you see? The answer from many would be "The Man in the Moon". The pattern feeding such imagination consists of the arrangement of light and dark patches. As noted in chapter 7 the early forming Moon was awash with molten lava eventually finding its own level, contracting and solidifying in and around rocky surface features. Areas resembling marshes, lakes, bays and seas fixed themselves amongst the hills, mountains and crater walls. Because of this likeness they were named palus, lacus, sinus and mare (plural maria). The largest are, of course, the maria and the largest of those provide the first-glance recognizable patterns. The following Full Moon photograph illustrates this (figure 8.1). The names of the major maria and the large ocean are labelled. Even the smallest of telescopes or binoculars reveal this first fascinating view of another world and it is well worth memorizing the maria names as an initial step to becoming familiar with features of the Moon.

Although shapes of the maria and other large lava areas are not perfect circles here are lists of approximate diameters or lengths to provide some idea of absolute and relative sizes (tables 8.1 and 8.2). For some maria different references give widely differing size values, probably reflecting the maria asymmetry and the orientation of measurement. The English equivalent names are also given, as many prefer to work with those.

Figure 8.1.

Table 8.1.

Mare	English name	Size (km)
Oceanus Procellarum	Ocean of Storms	2,568
Frigoris	Sea of Cold	1,350
Imbrium	Sea of Rains	1,300
Australe	Southern Sea	900
Tranquillitatis	Sea of Tranquillity	873
Fecunditatis	Sea of Fertility	840
Nubium	Sea of Clouds	750
Serenitatis	Sea of Serenity	650
Crisium	Sea of Crises	570
Humorum	Sea of Humours	400
Marginis	Marginal Sea	360
Smythii	Smyth's Sea	360
Nectaris	Sea of Nectar	350
Orientale	Eastern Sea	327
Vaporum	Sea of Vapors	245
Undarum	Sea of Waves	243
Humboldtianum	Humboldt's Sea	160
Spumans	Foaming Sea	139

And a few smaller lava fields commonly referred to.

Table 8.2.

Mare	English name	Size (km)
Palus Epidemiarum	Marsh of Epidemics	300
Palus Somnii	Marsh of Sleep	240
Palus Putredinis	Marsh of Decay	180
Palus Nebularum	Marsh of Mists	150
Sinus Roris	Bay of Dews	500
Sinus Medii	Central Bay	350
Sinus Aestuum	Bay of Heats	230
Sinus Iridum	Bay of Rainbows	200
Lacus Somniorum	Lake of Dreamers	230
Lacus Mortis	Lake of Death	150

The far side of the Moon contains no significant maria areas save Mare Orientale that sneaks onto both sides.

From dark patches to bright streaks: with the naked eye and especially through binoculars the enormous brilliant splat over much of the southern Moon dominates the vision. The meteorite that gouged out the very deep crater Tycho left

Figure 8.2.

this additional huge calling card – the bright radiating spokes of ejecta reaching areas hundreds of kilometers away. The debris covers so many of the other craters that it is clear that Tycho was formed after others in the area (figure 8.2). More on this later. Also prominent are the rays of Copernicus and Kepler. Proclus amongst the rocks surrounding Mare Crisium and Menelaus on the edge of Mare Serenity are associated with smaller but easily found rays. Many bright spots seen on the face of the Full Moon hint at the multitude of lesser ray systems.

Three more features are well known, and easily picked out by a keen eye or binoculars (figure 8.3). Aristarchus is by far the brightest crater on the Moon and its brilliant spot stands out at all times that it has any chance of illumination, even at Earthshine. More about this interesting area later. Grimaldi, seen towards the western limb, is a huge crater or lava plain with a dark floor. To the north, the

Figure 8.3.

circular and also dark-floored crater of Plato stands out like a large eye at the edge of Mare Imbrium.

Together, the maria, rays and other features described present a wonderful first introduction to the wonders still to be viewed and photographed in more detail.

Now, how on Earth can the features of the Moon be remembered with so many craters, dark areas, light areas, mountains and lots of other bumps and cracks! First thing is to have a good idea of which way up the Moon is and position the main features. Then it makes the next stage of learning, hard work and familiarization, a great deal easier. We need a starter vision. Three sketches will suffice.

Many describe the pattern on the Moon as seen by the naked eye as a man, a rabbit, a witch and others. Ever thought of a cow's udder? Draw such a shape within a circle (figure 8.4).

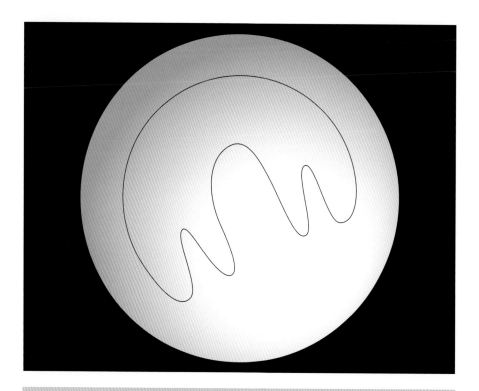

Figure 8.4.

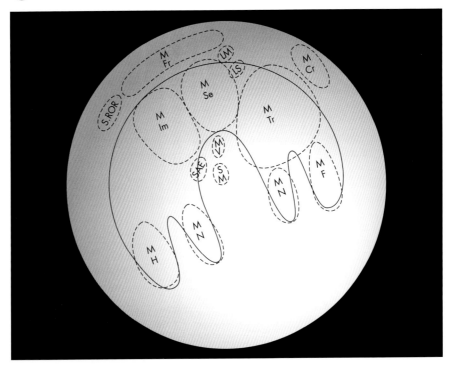

Figure 8.5.

At each "prominence" draw a circle (figure 8.5). Each circle represents a sea or mare. Add a few other circles on the blueprint to represent major seas, lakes and bays – the initials in the diagram should be obvious but check with a lunar atlas if not.

Lastly, pinpoint approximate positions of the main craters and crater chains (figure 8.6). PAA is Ptolemaeus, Alphonsus and Arzachel, PRW is Purbach, Regiomontanus and Walter. Theophilus, Cyrillus and Catharina are TCC. Grimaldi, Gr, is almost opposite Mare Crisium. Have you noticed that the older astronomers/philosophers/scientists crater names are towards the north pole and the later ones towards the south?

Then, at your own pace, insert more features, maybe beginning with the major mountain ranges. Now just where was that beautiful crater, Mersenius, with the domed floor!

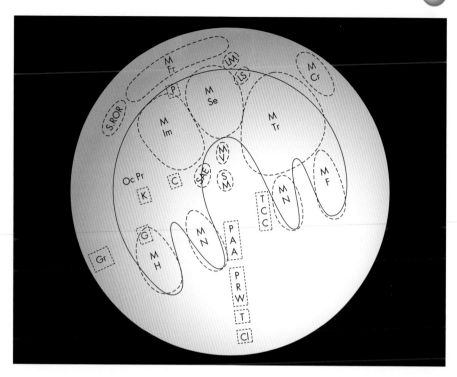

Figure 8.6.

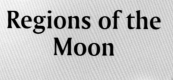

CHAPTER NINE

Regions of the Moon

Introduction

Now what to do with all those glorious pictures? How do you know what they are? There are many books and references to help, examples of which are given in the appendix. The commonly used *Hatfield Lunar Atlas* divides the Moon into 16 areas so it seems reasonable to select one of your pictures for each of those sections (figure 9.1). Consult this Atlas or a similar reference to help with identifying features within the images. The diagram shows the approximate regions corresponding to each area. Note also that the orientation of the Moon with respect to its north–south axis is shown as actually seen, also known as binocular view. The following pictures are examples taken from each of the 16 areas obtained using the described technique. Each picture is accompanied by an annotated reduced image to pinpoint features mentioned in the text.

1 Mare Serenitatis Region

The Menelaus crater (1) at the edge of the Montes Haemus (2) is 26 km across and is the origin of a small ray system just visible but more prominent at other conditions of illumination (figures 9.2 and 9.3). Although relatively small it is one of the brilliant craters easily picked out at most angles of sunlight. Its walls are around 2,400 meters high. The Montes Haemus themselves can present striking illumination and, with a few breaks from valleys and "lakes", stretch 560 km along the border between the Maria Serenitatis (3) and Vaporum (4). Bessel (7),

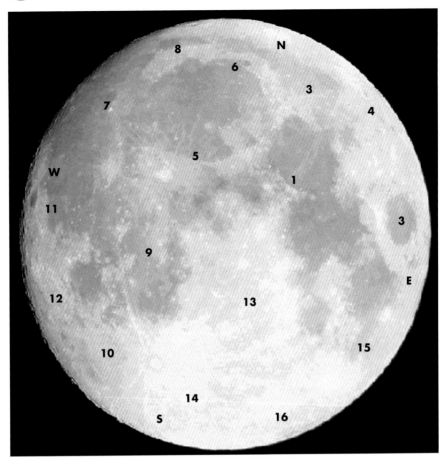

Figure 9.1.

with walls half the height of Menelaus, is the largest crater within Mare Serenitatis at 15 km in diameter. A bright ray of Menelaus passes close by, once thought to emanate from the distant Tycho. Sulpicius Gallus (12 km) (5) lying to the west of Bessel and Menelaus with walls up to 2,700 meters, is a bowl-shaped bright crater nestling near the base of the Montes Haemus. Julius Caesar (6) to the south has an incomplete wall system, dark floor and is 90 km across. Quite noticeable in this picture are the results of the catastrophic effect of formation of Mare Imbrium following the impact of a huge asteroid or similar cosmic visitor. The "scratch marks" radiating from Mare Imbrium (top left of the picture) can be seen everywhere but particularly across the ranges of mountains. The devastation of some crater walls, for example Julius Caesar, also testifies to the monstrous explosion. Some of the darkest regolith of the Moon is to be found here, to the south of Mare Vaporum, probably caused by explosive volcanic ash covering a lump of ejecta from the Mare Imbrium impact. As with most Moon pictures there

Figure 9.2.

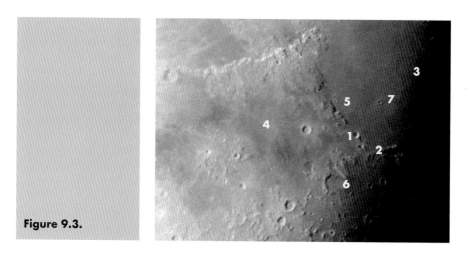

Figure 9.3.

is so much more detail to bring out – see later as to how. Remember that all images here are represented as binocular vision, that is, just as you see the Moon with the naked eye. North–south is designated top–bottom and right–left is east–west.

2 Mare Frigoris Region

Mare Frigoris (1), at 1,596 km long, provides an irregular shaped partition between the 261 km Alpine mountain range and the north polar uplands (figures 9.4 and 9.5). In the picture the camera has brought out the fascinating texture of the terrain around the two dominant craters of Aristoteles (2) (87 km) with its radiating spokes of ridges and Eudoxus (3) (67 km). Both have high walls of over 3,000 meters. Aristoteles has been described not only as a crater but also a great plain since the floor is flat having been flooded early in its history by lava. The flood covered all but a few mountain peaks within. To stress the advantage of filming features close to the terminator, lovely shadows are captured here that clearly show terracing inside the walls. Just outside Aristoteles and part of its wall to the east, is a smaller but deep crater Mitchell (4) at 30 km. Clearly the good atmospheric seeing conditions and low angle of the Sun have brought out excellent surface detail. At the terminator itself the rugged nature of the surface is picked out beautifully. Many small craters can be seen, such as Lamech (5) at

Figure 9.4.

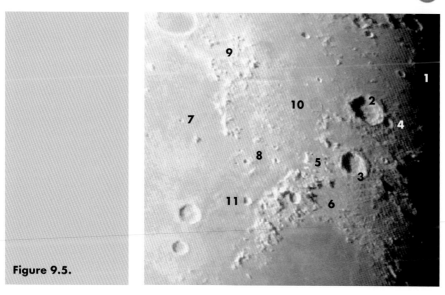

Figure 9.5.

13 km but very gratifying is the capture of features with low and broken-down edges such as Alexander (6) at 82 km, which is more recognizable as a small dark plain rather than a crater because of the disintegration of the walls. About 80 km to the west of Aristoteles is an unusual small square of straight hills. It is in fact the crater Egede (10) at 37 km. Theaetetus (11), 25 km across, is an example of how deceptive some measurements can be. Its floor is around 2,300 meters below the rim – quite high. However, the floor itself is nearly 2,000 meters below the surrounding surface so that the crater walls appear very low from the outside at only a few hundred meters. Many tiny hills abound in the area. Cassini (8), surrounded by its irregular walls and ridges and double crater within, is a very distinctive feature 57 km across. Is that a crater larger than and between Aristoteles and Eudoxus or is it an accident of formation in that area?

Mare Frigoris is very irregular in shape and has an area of 441,000 km^2. It may have been an overflow of lava during early days of formation. Via "lakes", "bays" and a valley it has links with Maria Serenitatis, Imbrium (7) and Oceanus Procellarum. The Montes Alpes (9) and the photogenic Montes Jura provide its southern border.

3 Mare Tranquillitatis Region

Prominent in this sector is Mare Crisium (1) of diameter 570 km. Its dark floor helps to highlight the scrumptious detail of the surrounding highlands. Brilliant Proclus (2) (28 km) reveals its rays and Mare Crisium's surround on the edge of the terminator shows its rugged nature (figures 9.6 and 9.7). The terminator, the line of the Moon that just enters shadow, cuts right across Mare Crisium in this image.

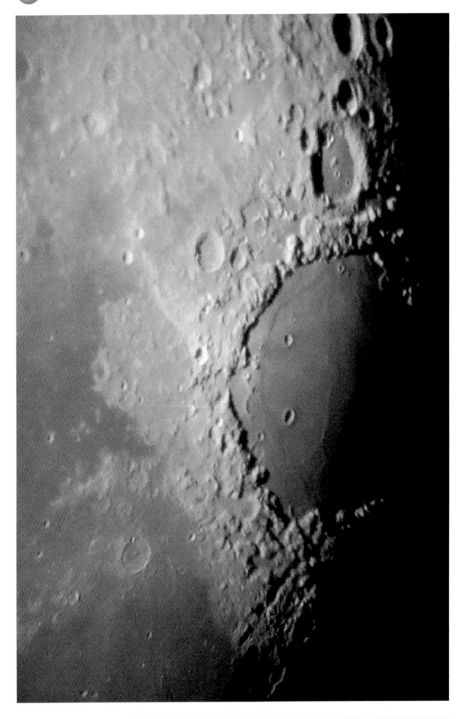

Figure 9.6.

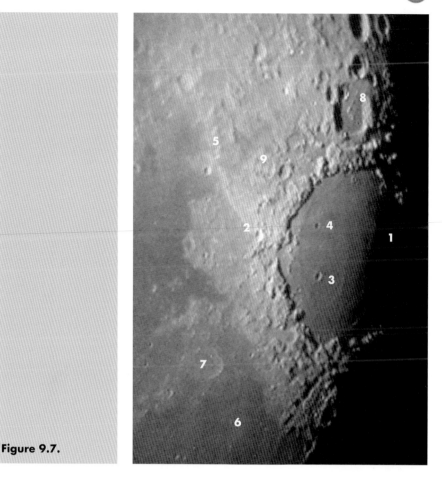

Figure 9.7.

As this image is viewed Mare Crisium appears to be much longer from top to bottom than from side to side, which emphasizes the illusion of "foreshortening" towards the edges of the Moon (and any other globe). The mare actually measures longer east to west at 620 km than north to south at 570 km. It appears at New Moon until four days old and is visible with the naked eye as a dent at the edge. It does not actually spread far enough to appear on the far side of the Moon; of the major mare only Mare Orientale has that distinction. Mare Crisium's ease of identification is partly due to the fact that it is clearly defined by not having any connections with other maria. If Mare Crisium is viewed at New Moon, small black dots of craters are seen punctuating the large flat lava plain. However, best details are picked out around day 16 as illustrated in the image above. The northern, western and particularly the southern mountain borders are wondrous in their complexity and ruggedness beautifully brought out at this angle of illumination. The picture shown in figure 9.8 also shows the rugged eastern border where one of the highest features of the Moon can be found, Cape Agarum at around 5,500 meters. In 1953 John O'Neill announced that he had found a huge natural arch crossing from what were then called the Promontaria of Olivium and

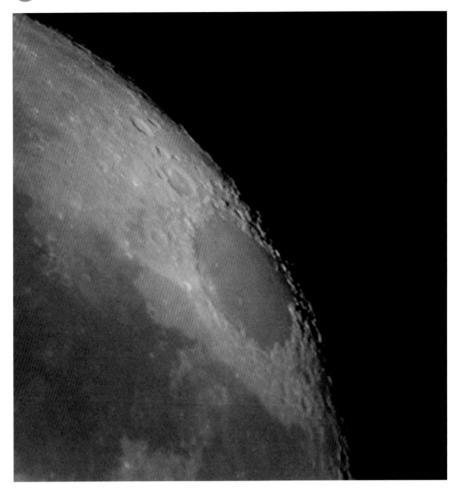

Figure 9.8.

Lavinium. Following much debate and speculation, mainly by the amateur astronomer Percy Wilkins, it was eventually proved to be a shadow effect and all thoughts of an artificial or natural bridge were rejected. Crater Picard (3) at 23 km is the largest within Mare Crisium with high walls of 2,700 meters and Peirce (4), at 19 km and walls of 2,300 meters, a little smaller. The craters of Swift (11 km) next to Peirce, and Greaves (14 km) next to Picard can be seen, as can the broken-down craters of Lick (31 km) and Yerkes (36 km) in the same vicinity. Proclus is the star of the show by exhibiting a very bright ray system squirting its arm out to Carmichael (20 km) and, it appears, every other direction except toward the land mass Palus Somnii directly in front to the west. It is one of the brightest points of the Moon and with walls 2,700 meters high. The terminator, where most images of interest are to be found, cuts right across Mare Crisium in this photograph. Also in the picture many tiny craters are easily visible in Mare

Fecunditatis. Anville (6) is only 16 km across and close by Smithson at just 6 km. Taruntius (56 km) (7) is an unusual crater in having a somewhat concentric crater within. Its walls are low at just 1000 meters. It also has a central mountain with a dent in the top. The wrinkles and folds in the solidified lava plain of Mare Crisium are well-defined as lit by the low evening Sun.

Just north of Mare Crisium lies the glorious crater Cleomedes (8). At 126 km long it is shown to great advantage with its high walls and peaks rising to 3,000 meters and more. Several smaller craters reside within, including the very

Figure 9.9.

deep Tralles, 43 km in diameter. Almost touching Cleomedes is a small complex group around Burkhardt (57 km) and further out, Geminus (85 km) with its high walls up to 5,000 meters in places next to Bernouilli (47 km).

This view of opposite illumination (figure 9.9) highlights the craters within the lava plain and, just in the darkness of Mare Tranquillitatis, Taruntius reveals some inner detail.

The shadows within Macrobius (9) caused by its wide crater of 64 km and very high walls up to 4,000 meters make it a most conspicuous feature of the moonscape.

Mare Tranquillitatis is one of the major maria and somewhat lighter in color than others. It is connected to other maria – Serenitatis, Nectaris and Fecunditatis with Crisium and Vaporum nearby. The famous Apollo, first men on the Moon, landing took place here but more on that later. Mare Tranquillitatis is packed with interesting features that require low-angle lighting and close inspection to identify. Some overexposure was necessary to capture many of them in this western portion of the mare (figures 9.10 and 9.11). Immediately obvious, of

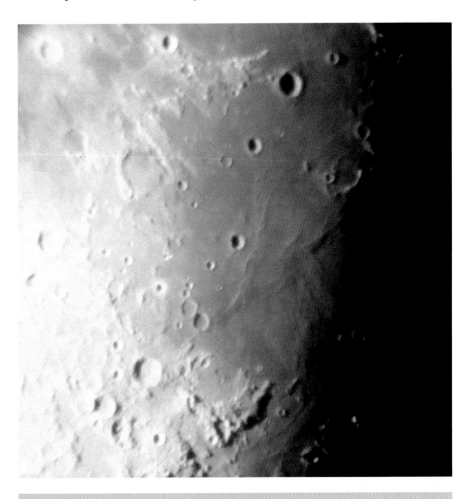

Figure 9.10.

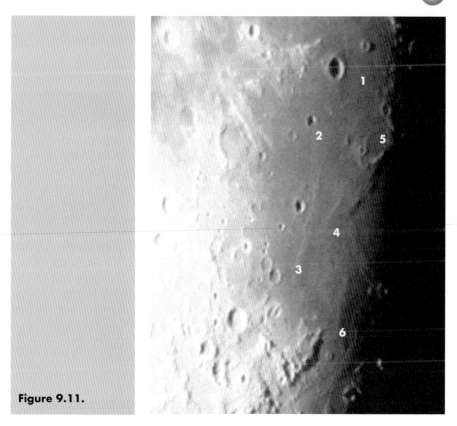

Figure 9.11.

course, are the craters. Plinius (1) at the border with Mare Serenitatis is deep at 2,320 meters, 43 km across and has high terraced walls. It is usually apparent under most angles of illumination. Much less conspicuous is Maclear (2), a dark floored smaller crater at 20 km in diameter, with Ross (24 km) nearby. To the south (3) are two pairs of craters – the larger Sabine (30 km) and Ritter (29 km) and two craterlets. Carrel (5) (15 km) sits amid ridges and close to a short and narrow range of hills. Arago at 26 km has an irregular shaped rim. The landscape (marescape?) owes much of its beauty to the wrinkles formed as the lava cooled and compressed some 3.5 billion years ago. Careful inspection is required to discern the very low wrinkle edges of Lamont (4) over a buried crater. It has a diameter of 106 km. It engendered much interest when it was discovered that it was the source of a mascon (mass concentration), a particularly dense region of the surface. Mascons had been noted to have an effect on the motion of circulating lunar orbiters. Close by is Arago especially noted for its proximity to small lumps in the surface lava nearby. They are referred to as domes and the two north and west of Arago can just be picked out at advantageous lighting conditions. The domes are known as Arago alpha, to the north, and Arago beta to the west. Domes are only a few hundred meters high but several kilometers across and therefore appear as gentle swellings or surface upheavals. Because of their smooth and featureless surface they are tricky to observe unless close to the terminator under low-angle sunlight. Between the low-walled, irregular Hypatia

(40 km) and the complex structured Torricelli (6) (22 km) just within Mare Nectaris several craterlets and wrinkles are nicely highlighted. Apollo 11 landed near here – more on that later.

4 North-East Region

While the Sun's light all but obliterates many of the features in this image (figures 9.12 and 9.13), Atlas (1) (87 km) and Hercules (2) (69 km) can be seen leading on to Endymion (3) (123 km) at the terminator with its high walls (4,500 meters) and dark floor. It is a very circular walled plain and larger than similar craters such as the more frequently recognized Plato. The image has caught Endymion nicely at a low angle of illumination to reveal two more large walled areas. The adjacent structure surprisingly does not have a name but immediately beyond lies De La Rue. At 134 km in diameter it also has low and broken-down walls. Strabo (55 km) sits at the edge of De La Rue and is the source of a minor ray system while Thales just beyond at 31 km is a major source of rays especially visible at Full Moon. Consult a Moon atlas for complete identification of craters in this patch.

To the south of Endymion lies the oblong Messala, a huge enclosed plain at 125 km with low and irregular walls. Close to Messala is the smaller Hooke at 37 km, an impact crater filled with lava to produce a dark flat floor. A little further in, conspicuous Cepheus (39 km) with a small bright crater within its wall pairs with Franklin (56 km).

Figure 9.12.

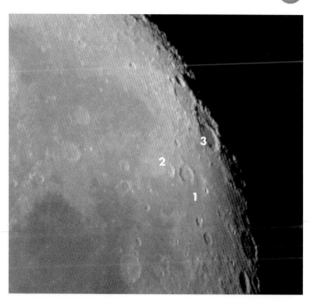

Figure 9.13.

5 Mare Imbrium Region

Well-known Copernicus (1) (107 km) and Eratosthenes (2) (58 km) are highlights here (figures 9.14 and 9.15). More of these two magnificent craters later. Note the tiny craters such as Draper (3) (8 km) easily captured using this standard equipment. How rewarding to see the beautiful surface coloration from ray systems and the sunlit hills, Montes Apenninus. Remember, of course, that the color contrasts are in reality shades of gray but the overlay of browns caused by the light passing through our atmosphere provides an additional beauty. The detail is particularly satisfying considering the inexpensive equipment used. Archimedes (4) at 83 km in diameter is of regular shape with a dark flat floor (scattered with a few craterlets) and the largest crater in Mare Imbrium. Its walls have been eroded and are not high, although over 2,000 meters is attained in places. Nearby Bancroft at 13 km is not huge but is a very deep 2,500 meters, especially for its size. Two tiny craters, Beer (10 km) and Feuillée (9 km) (5), are not far away to the west before coming to Timocharis (6). Timocharis is 33 km in diameter although not very circular and the source of a minor ray system. It appears to have a central peak that is, in fact, a central crater. What a good shot by that meteorite! Well-defined terracing is also within, along with many small radial ridges. Lambert (7) at 30 km, a much less bright crater, lies still further to the west. This, too, sports a central crater. A lava ridge running right through Lambert towards Eratosthenes can just be seen in this image although more prominent at sharper angles of illumination. Pytheas (8) at 20 km is very bright for its size. Many tiny features of mountains and craters are to be found in the lava sheet such as Heinrich at only 6 km across and Huxley at 4 km. Once more the potential of the resolution achievable with the described equipment is made clear in this image since even smaller craters and features, although unnamed in standard Moon atlases, can be picked out. The Montes Apenninus with their rugged formation are a wonderful sight in the "evening" sunlight. They stretch for 401 km and contain the highest peak at a

Figure 9.14.

lofty 6,000 meters. A small ribbon of mountains lie in front of Mons Ampère (9), which is 30 km across and around 4,000 meters high.

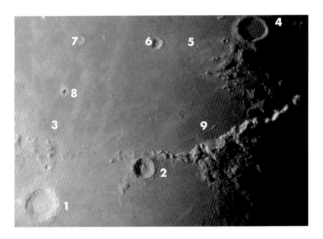

Figure 9.15.

6 Plato Region

Plato (1) must be among the most recognizable of craters with its 100 km diameter, dark flat floor and regular walls (figures 9.16 and 9.17). Much detail can be seen across the Mare Frigoris (2) towards the terminator as well as the small, pic-

Figure 9.16.

turesque mountains. Mons Pico (3) is a triple-peak bright mountain range just 25 km long but over 2,500 meters high. Mare Imbrium (4) is the largest of the regular shaped seas at a diameter of 1,123 km and hosts many beautiful sights – mountains, craters, and bays.

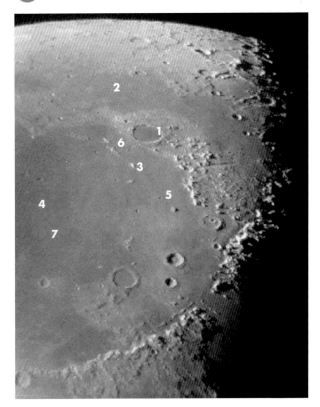

Figure 9.17.

Several small craters are easily noticed in the smooth flat lava. Piazzi Smyth (5) at 13 km and nearby Kirch at 11 km are two bright ones near another of the isolated mountains, Mons Piton, which is over 2,000 meters high and has a small crater at its peak. The Montes Teneriffe (6) are a cluster of small bright peaks rising to 2,400 meters. A large buried or destroyed crater is thought to exist in the area bounded by Plato and the mountains of Teneriffe and Pico. The ring is accentuated by the lava wrinkles and was unofficially named Ancient Newton. It is often referred to as the ghost ring. The name of Bliss has now superseded that of Ancient Newton following a suggestion by Patrick Moore. The tiny Landsteiner (7) at 6 km can just be seen in this image. The grand Archimedes appears once more in this sector with its very regular circular crater at 83 km in diameter, walls around 1,400 meters high and a dark, flat, slightly sunken floor. The image has captured the streaks of a ray originating from Autolycus across its floor.

7 Oceanus Procellarum

A large mare on the west of the Moon (figures 9.18 and 9.19). In the lower center is Kepler (1) (31 km) a source of one of the major ray systems as well as the Schröter valley (2), shown in the next sector, and Sinus Iridum (3). More details

Figure 9.18.

on these features later. Three craters stand out to the north of Oceanus Procellarum since they are bright and clear cut. Bianchini, set prominently into the Jura mountains (4), is 38 km in diameter as is Sharp at 39 km. Both are terraced within, have flat floors and small central peaks. Mairan is similar at 40 km but without a central peak. Beyond Montes Jura are two large lumps of rock about 1,000 meters high and 20 km across, making up Mons Gruithuisen (5) named after the very first proponent of the modern theory of lunar crater formation from falling meteorites. Franz von Paula Gruithuisen put forward his theory as long ago as 1824 but it was put aside and forgotten until much later. A small crater also named after him lies just to the south having a diameter of just 16 km.

Much of the beauty of the huge expanse of the lava plain that is Oceanus Procellarum lies in the many small craters picked out amongst the ever-changing delicate pastel colors. Caroline Herschel (1,850 meters deep) at 13 km and nearby Heis at 14 km are two easily found bright craters. In the same area and somewhat brighter and larger is the pair Delisle and Diophantus at 25 and 17 km both having central peaks. Mons Delisle with its peculiar tooth shape lies close by.

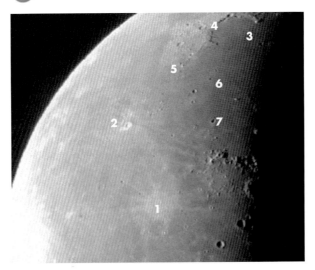

Figure 9.19.

Domes are to be found in this area although at the magnification used to obtain this image they are not easy – but not impossible – to spot. Euler (27 km) is prominent, bright and the source of a small ray system. Bessarion, a small bright crater at just 10 km is found on the way to Kepler. Of course Kepler is very well-known as the origin of the third most prominent ray system, although Kepler crater itself is a very modest 31 km with low walls and little relief. However, it is quite deep at 2,570 meters. Amongst one of the ray spokes reaching to Brayley via Bessarion small domes can just be picked out. A satisfying photographic capture of a lava ridge can be seen at the terminator stretching from the Promontorium Laplace in the Jura Mountains and winding its way through to Lambert via C. Herschel. It is well worth taking time to locate and identify the many craters and mountains in Oceanus Procellarum from a standard lunar atlas. The spectacular Vallis Schröter is described later.

8 North-West Region

The dominant feature here is the Schröter valley (1) (figures 9.20 and 9.21). Different illuminations produce very different pictures that make it worth photographing at every opportunity. It winds for about 200 km away from Herodotus (2) (34 km) and the brilliant Aristarchus (3) (40 km). More on this later. The moonscape in this region is pockmarked with small craters. Krieger (4) (22 km) is an interesting shape and with its broken-down walls is host to the tiny Van Biesbroeck at just 9 km within its ramparts. Angström (9 km) and Nielson (10 km) can also be found in this area. The Harbinger mountains (5) are a scattered clump of hills rising to 2,500 meters in places. They may have been part of the border or crater wall of Imbrium according to current belief that Imbrium is, in fact, a gigantic crater since the floor is lower than surrounding areas. Once

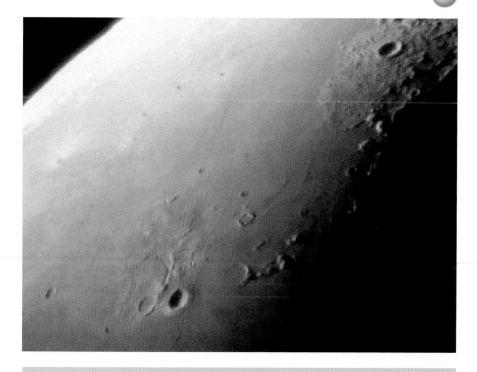

Figure 9.20.

Figure 9.21.

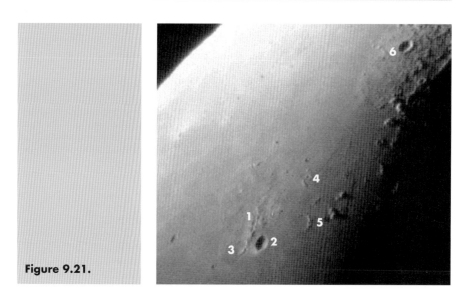

again the angle of illumination allows the picturesque lava ridges to be seen stretching from the Schröter valley to the Jura mountains where the clear highlit profile of Mairan (6) (40 km) can be seen.

9 Mare Nubium Region

A busy section of the Moon indeed with many interesting features some of which are discussed later (figures 9.22 amd 9.23). Mare Nubium is one of the larger maria at approximately 715 km diameter. It has many small features and its borders are less regular than some other maria such as Mare Imbrium. Two craters showing brightly are the regular shaped Lalande (1) and Mösting (2) both 24 km and with a small central hill. The latter is often used as a reference point for locating other features. It is also quite deep at 2,760 meters. The Fra Mauro (3) (101 km) group contains an interesting "finger" shape emanating from Guericke (4) (63 km) but more on this group later. There is an interesting formation to the east of Fra Mauro across a part of Nubium via the small but bright Kundt (10 km). It is a large and so far unnamed lava plain (5). Prominent on its southern boundary is the high-walled Davy (35 km) that also hosts the deep small crater Davy A in its own walls. To the north of the plain is Palisa at 35 km and within the remainder of the border lie several small craters and elevations. Just to

Figure 9.22.

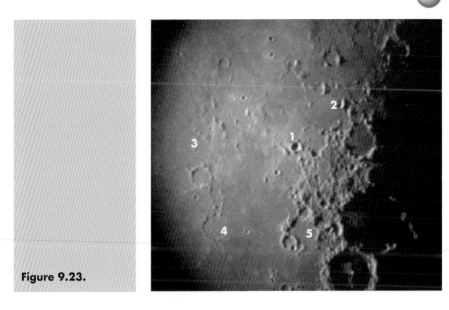

Figure 9.23.

the north of Palisa is a very recognizable "V" shape formed by two lines of craterlets, one reaching to Lalande and the other to a small "cove" nearby. Just north of Mösting are two broken walled craters, Sömmering at 28 km in diameter and the larger Schröter at 35 km that is nowhere near the Schröter valley.

10 Southern Moon – Tycho Region

The southern Moon is absolutely full of craters. Tycho (1) (85 km) is probably the best known and is the origin of the largest ray system (figures 9. 24 and 9.25). The bright rays totally dominate the landscape of the Full Moon being the most noticeable feature seen by the unaided eye. The projectile that carved out the formation of Tycho must have caused an incredible explosion. The crater is 4,850 meters deep with much internal structure such as central peaks up to 1,600 meters. The crater came into existence towards the end of any active bobardment of the Moon, possibly 100 million years ago, as evidenced by the rays spreading across the surface of other craters and features for huge distances. The ejecta forming the rays is not just a dusting of powder but several meters deep in places. Deslandres (2) is a very large and complex crater with others inside – notably Hell (33 km). Deslandres is magnificent in its vastness spanning a huge 256 km in diameter. It is old and has had a hard life with its crumbling and crater-punctuated walls. Lexell at 63 km sits on its southern wall and has itself suffered demolition of its northern edge to allow an inflow of lava. A little further on from Lexell is the high-walled Ball at 41 km. Hell is a deep (2,200 meters) and regular crater. Many smaller craters lie within and the myriad bumps and hills give the floor a very rugged appearance. Of much interest is the speculation that amongst the internal features can be discerned outlines of what are possibly several large but destroyed or buried craters or plains adding to the notion that

Figure 9.24.

Figure 9.25.

Deslandres is of great age, possibly close to 4 billion years. Between Tycho and Deslandres lies another ancient and much eroded crater, Sasserides at 90 km across. Again, the walls show later craters and the floor contains many dents and hills. To the north-west of Tycho lies the medium sized Heinsius at 64 km whose southern wall is severely disrupted by more craters. Longomontanus (3) is also large at 145 km its ancient age being characterized by eroded walls and much detail within.

11 Mare Humorum and South Oceanus Procellarum

Gassendi (1) (110 km) is prominent in this area (figures 9. 26 and 9.27) and very recognizable from its circular shape and sharp outline. Although the walls appear shallow they rise to 3,600 meters in places but low, almost to the floor of Mare Humorum, in others. Gassendi A (33 km) cuts right into its northern border with Gassendi B beyond at 25 km. A triangle of shadow is apparent at this point

Figure 9.26.

Figure 9.27.

when near the terminator. Its base holds several fairly high peaks, one up to 1,000 meters. Apart from bumps and mounds the floor is host to many cracks and ridges. It is therefore not surprising that many transient (short-lived active) phenomena have been sighted here. These "TLPs" are described later. Just about every area and feature of the Moon seems to be picturesque, attractive or unique in some way. A trio of individually distinctive craters merge at the southern end of Humorum (2). In the center is Lee at 41 km, although it is more of a bay than a crater; it may even be two joined craters with destroyed walls and flooded with lava. Doppelmayer at 64 km is quite circular although the northern wall has been demolished. A prominent peak stands out nicely in the center accompanied by other smaller hills. Vitello is an approximately concentric crater with a central peak giving home to a craterlet at the peak. Mare Humorum is relatively featureless but shows several small craters and lava ridges to the east. A significant scarp is Rupes Liebig around 180 km long and, although very low in places, leads right up to Gassendi. It passes by Liebig crater (2) itself (37 km) and De Gasparis (30 km). The Riphaean Mountains (3) (189 km long) are just visible in the sunlight to the north.

12 South-West Region

Schiller (1) is a combination of two or possibly three impact craters. It is considerably longer at 179 km than wide at about 70 km. The long flat plain has some mounds and ridges on its surface. Schiller is a very easily recognizable feature of the area (figures 9.28 and 9.29). Just towards the edge of the Moon from Schiller

Figure 9.28.

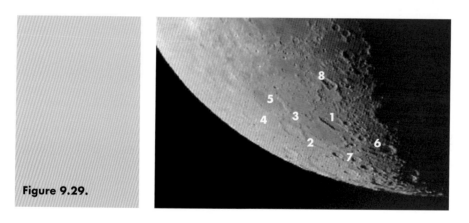

Figure 9.29.

lies a triplet of craters – Phocylides (2) (121 km), Nasmyth (3) (76 km) and Wargentin (4) (84 km). The latter has been completely filled by lava to form a smooth plateau. More on these later. Schickard (5), at 227 km, is one of the largest craters. Its walls are not high, at around 1,600 meters. The floor has much variation in color and contains many small craters and hills, the highest rising to about 2,500 meters. On its inner wall sits reasonably symmetrical Bayer (47 km) with its high terraced walls. Beyond Schiller towards the terminator sits the small-ish (compared to some of the giants in the area) Rost at 49 km then Scheiner (6) at 110 km with its companion Blancanus that is hidden in the shadow of the terminator. Towards the Moon's limb here lies a group of five large craters although only three are prominent in this picture. Zucchius (7) (64 km) and Bettinus (71 km) have high, continuous and terraced walls while Kircher (73 km) has little in the way of features on its flat floor. Hainzel (8) (70 km) is an interestingly shaped merger of two or possibly three craters forming a triangle and having peaks and undulations within. Certainly looks like an extraterrestrial to me! They border a large and mostly destroyed Mee (132 km) on the limb side.

13 South-East Central Region

This area (figures 9.30 and 9.31) is packed with well-known features – large craters with much structure inside such as Walter (1) (128 km), Arzachel (2) (96 km) and Albategnius (3) (114 km). More later on this group and other major features nearby. It is important to concentrate on the shadow opportunities provided at the terminator. Four good-sized craters show themselves – Aliacensis (4), Werner (5), Apianus (6) and Playfair (7). Aliacensis, 80 km in diameter, can be seen to have much terracing on the inner eastern wall while the other sides are severely damaged. It has a small peak in the center. Its surrounds show much

Figure 9.30.

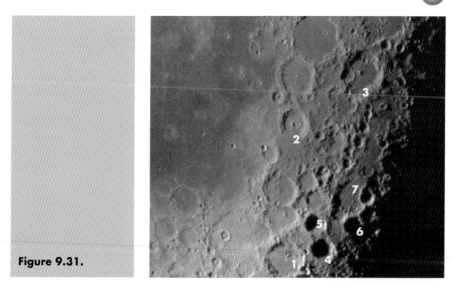

Figure 9.31.

scoring, dents and hills – quite a rough terrain. Werner is 70 km across and accompanies Aliacensis amongst the rough ground. It is a very well-formed and preserved crater with high, neat and terraced walls rising to 5,000 meters. The shadow of its high central peak can just be picked out in this image. Apianus (63 km) has high, terraced walls interrupted in places by damage and craterlets. Joining it via a wall/hill system is Playfair at 48 km. Apianus and Playfair encroach onto the walls of a large broken enclosure about 110 km across that contains a great many peaks nicely picked out by the evening Sun. A lovely string of small craters thread their way towards Albategnius – Faye (37 km), Donati (36 km), Airy (37 km), Argelander (34 km) and Vogel (27 km). The central peaks of Airy and Donati are picked out well in the sunlight.

14 South-East Region

This large expanse of the southern Moon (figures 9.32 and 9.33) is completely devoid of maria – just peppered with a myriad unique and beautiful craters and an overlay of the streaks of ejecta from the more recent Tycho that can be seen affecting the whole area. Hommel (1) (126 km) is shown off nicely at the terminator hosting three large craters, two within and one sitting on the wall. One has an easily visible central peak. Pitiscus (2) at 82 km in diameter is a fairly regular crater with high, but narrow, walls rising to over 3,000 meters in places. Much structure is visible – internal peaks and hills – and one very prominent crater. A portion of the Rupes Altai (scarp) can be seen that meanders for 480 km. It is thought to be the remnant outlier of surface movement following a giant asteroid impact that formed Mare Nectaris. The scarp leads right up to the very recognizable crater Piccolomini (3) (87 km) with high terraced walls containing peaks up to 5,000 meters and an unusual ingress at its southern edge. It has a huge mountain in the center at 2,000 meters. Stiborius (4) at 44 km across has a nicely

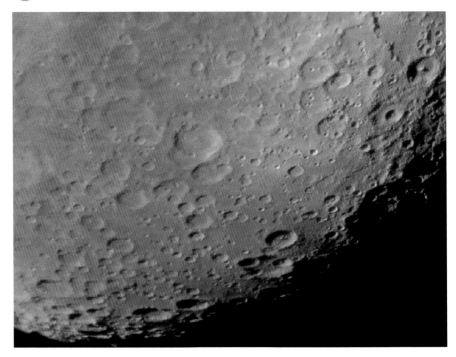

Figure 9.32.

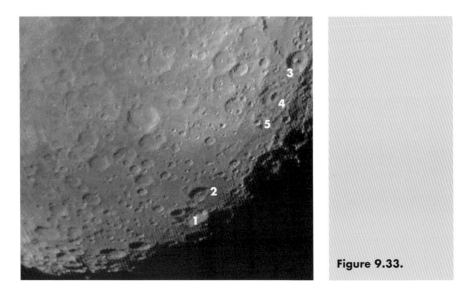

Figure 9.33.

visible, not quite central peak. It sits inside a much older and worn-down crater. Also sitting within an older but elongated plain is Wöhler (5) (27 km). Accompanying Wöhler in the oval appears to be a huge dome or rounded mountain. A reminder of the rugged nature of the surface can be seen right on the edge of the terminator with the strewn rocks, cracks and lines of tiny craterlets.

15 Maria Fecunditatis and Nectaris

The image highlights the intricate shapes of the craters and mountains bordering the Maria Nectaris and Fecunditatis (figures 9.34 and 9.35). Right in the center of this expanse is the Colombo (1) (76 km) Goclenius (2) (72 km) group of craters.

Figure 9.34.

Figure 9.35.

Gutenberg (3) (74 km) is an irregular crater with damaged walls, particularly so at the southern edge by a 20 km gouge. Several well-worn peaks reside within. Goclenius is paired with Gutenberg. Both are irregularly shaped and contain many small peaks and rills on their floors. Magelhaens (4) (41 km) is joined by a smaller crater to form a double. Colombo at 76 km in diameter is a very irregular sprawling enclosure damaged to the north-west by a crater about half its size. The Montes Pyrenæus (5) bordering Nectaris and pockmarked with many craterlets are a pretty sight. Langrenus (6) (127 km) with its massive walls can be picked out towards the Moon's edge. Also found near here (not shown) is the photogenic trio of craters Theophilus, Cyrillus and Catharina. More details of that trio later.

16 South-East Region

The crater Janssen (1) (199 km) is another that has had a hard life (figures 9.36 and 9.37). It has much structure on its floor and Fabricius (2) (78 km) is amongst several impacts to have broken down its northern walls. The bright crater Lockyer (3) at 34 km across sits on the ramparts of the southern wall. Because the walls are in such a poor state of repair it is really only well seen when at or close to the terminator. Regular Vlaq (4) at 89 km is shown beautifully in the picture so close to the terminator. The shadow has just bisected the deep floor to reveal the

Figure 9.36

full height of the very central peak. The walls rise to over 3,000 meters. The similar sized Rosenberger (95 km) is its joined neighbor with a smaller central peak.

Vallis Rheita (5) – the valley just north of Janssen – is in fact composed of a chain of craters stretching some 445 km. It is 50 km wide in places. It is not quite shown to its best advantage in this picture.

Figure 9.37.

CHAPTER TEN

Moon Features and Techniques

Be on the lookout for photo opportunities. Notice a small bright line in the shadow just to the west of Plato (figure 10.1)? Using a higher magnification eyepiece, view that area. Take a sequence of pictures – each time moving the telescope so that the camera spot-metering square is centered at intervals from fully onto the bright Moon to fully in the background shadow (figures 10.2 to 10.5). This will gradually increase exposure to bring out any detail.

The best image, which is one of the most spectacular of the Moon, can now be chosen for final display (figures 10.6 and 10.7). The "handle" is, in fact, the top of the Montes Jura (1) catching the Sun whilst the floor of Sinus Iridum (2) is still in darkness. Sinus Iridum at a diameter of 260 km is the result of an impact basin that has filled with lava. The Montes Jura that border the northern half of the bay rise to 2,000 meters while the southern half has disappeared under early volcanic activity. Wrinkle ridges at the front of the Bay attest to this. The result is that the light rays at sunrise skim the Mare Imbrium and Jura mountain peaks to create the jewelled handle image. Surrounding craters and rugged terrain bring out the beauty of the image captured. In addition to Helicon (3) (24 km) and Le Verrier (4) (20 km), the tiny craters of Landsteiner (5) (6 km) and McDonald (6) (7 km) indicate the resolution possible with the described simple equipment. The smallest named crater so far captured using this technique is Linné (2 km) in nearby Mare Serenitatis but much smaller may be attainable. Although small, Linné is easily found due to the larger light patch of surrounding ejecta.

Plato (101 km) is an excellent target for the telescope with its symmetrical shape and dark, flat floor. It is possibly the most well-known crater on the Moon and visible with the naked eye. It lies on the edge of Mare Imbrium, a beautiful sea containing a colorful (many shades of gray in actuality) flat surface dotted with tiny craters, domes and Copernicus and skirted by majestic mountain

Figure 10.1.

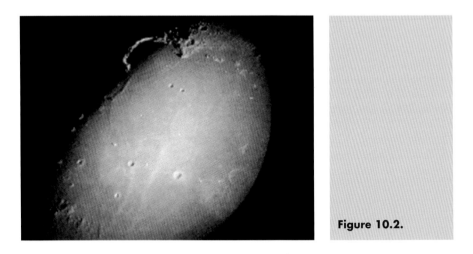

Figure 10.2.

ranges. Closer inspection of Plato reveals a pointed shadow being cast onto its floor at certain times indicating a mountain within the walls. A challenge is, therefore, to capture different shadow lengths such as the following: (figures 10.8 to 10.12).

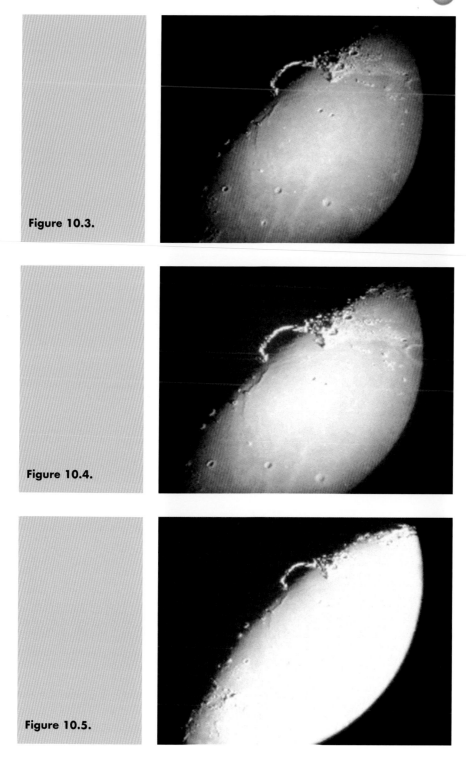

Figure 10.3.

Figure 10.4.

Figure 10.5.

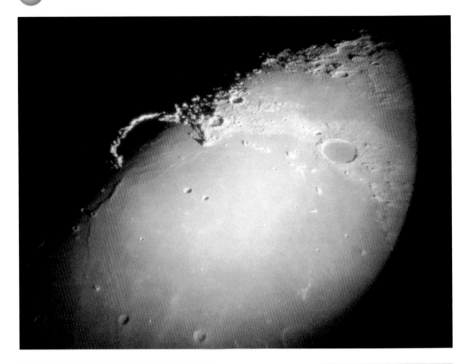

Figure 10.6.

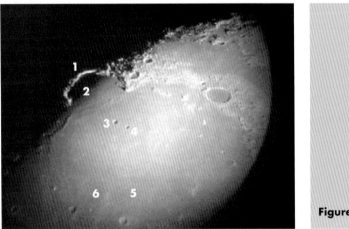

Figure 10.7.

First, no shadow. Then larger shadows. Then small and large shadows from the opposite side – note the light side of the Moon on opposite sides of the terminator, i.e. Plato's morning or evening. Finally the following best picture can be chosen for display.

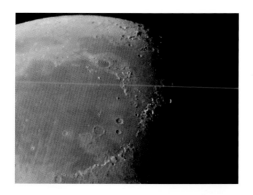
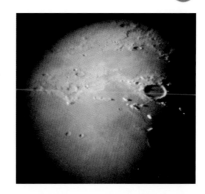

Figure 10.8.

Figure 10.9.

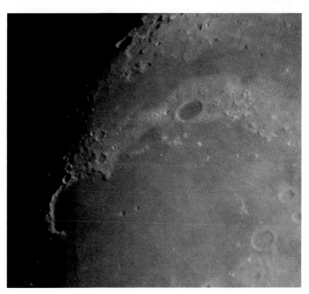

Figure 10.10.

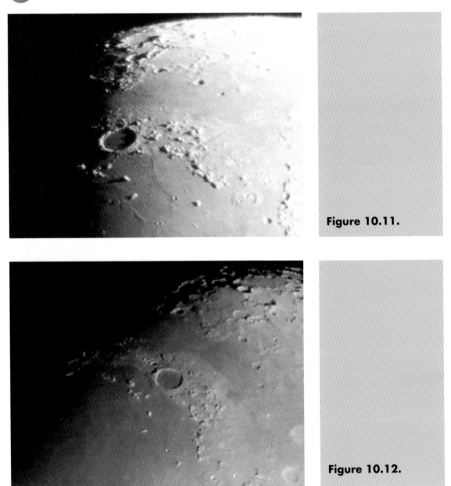

Figure 10.11.

Figure 10.12.

Note the detail within this image (figures 10.13 and 10.14). It has been captured at a time of extremely good seeing conditions and favorable lighting. Ripples and mountains (Mons Pico and Montes Teneriffe) to the south of Plato (1) form something of a reflection, referred to as Ancient Newton, that may be all that remains of a buried crater. As mentioned previously, Ancient Newton is now formally designated Bliss since another crater near the Moon's southern limb has the title Newton. The smooth flat floor of Plato contrasts sharply with the complexity of its walls and immediate surrounds. Well-defined long cuts can be seen in the outer slopes and others within the walls themselves. The deep and bright Plato A sits just outside to the north-west. The rugged terrain of the Montes Alpes reaching from Plato is beautifully illuminated. Nearby Piazzi Smyth and Kirch (4) at about 12 km each are picked out optimally at the terminator indicating the resolution possible with this instrumentation. Travelling just north from Plato across the wrinkled strait of Mare Frigoris lies a heavily eroded area. Birmingham (5) at 98 km across is a large walled plain whose walls have crumbled so much

Figure 10.13.

Figure 10.14.

that it can only be seen in satisfactory detail at sharp angles of illumination close to the terminator. The rubble-strewn floor is nicely picked out. Nearby Epigenes at 55 km is, by contrast, a well-defined crater with a smooth floor. Further out towards the limb, Anaxagoras (3) (50 km), which is the source of a major ray system, impinges on the much larger plain of Goldschmidt (125 km) with its low and broken walls. Because the libration is very favorable here another but unnamed large plain can be seen in much detail. Rocks and boulders are to be found over the floor surrounded by highly eroded walls. Even further out towards the limb is the raised plain (81 km) of Mouchez not referred to in all references possibly because a very favorable libration is required for it to be seen. Moving out from Plato across Frigoris once more can be seen the bright walled and distinct Fontenelle (6) at 38 km with a small central crater. Broken walled plains abound in the area creating a very picturesque scene. Sitting amongst them is the very deep Philolaus (2) (70 km) with heavily terraced walls rising to over 4,000 meters. The floor is cluttered with central peaks and hills.

Advantage must be taken at times of good libration and seeing conditions. Look out for shots such as this one (figure 10.15) showing grand Pythagoras (142 km) in its entirety at the limb with its high walls and massive central peak.

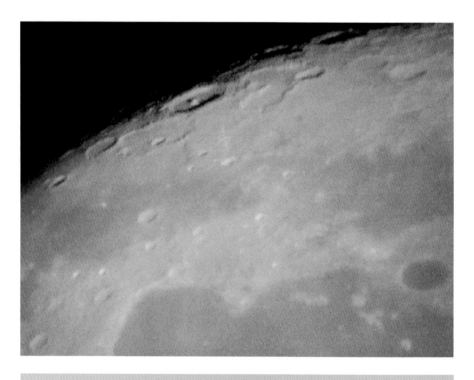

Figure 10.15.

More crumbled-walled plains surround Pythagoras, such as Babbage (143 km) with its inner crater. Is that Cremona, Desargues and Brianchon clearly visible at the edge? Wow!

Towards the east of Mare Frigoris are more highlights not least being the delicate lava ridges and wide range of color tones (figures 10.16 and 10.17). The face

Figure 10.16.

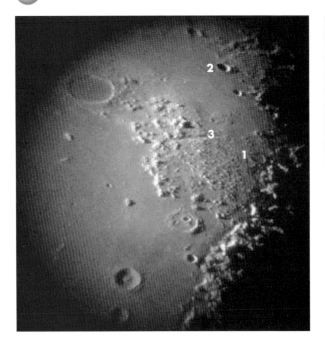

Figure 10.17.

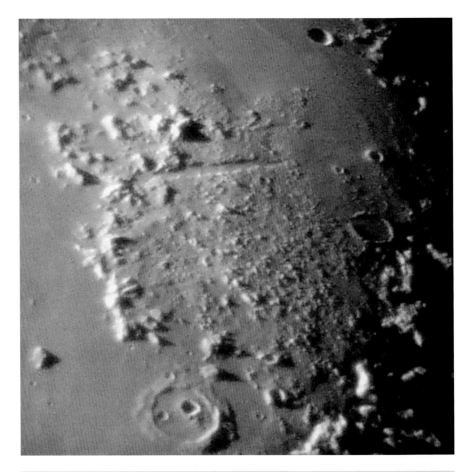

Figure 10.18.

of the Montes Alpes is brilliantly lit, casting short sharp shadows to show in high relief the tiny mounds and hillocks scattered towards Frigoris. Just beyond those scatterings and nestling at the foothills of the Montes Caucasus sits the strange rhombus-shaped Egede (1) at 37 km average distance across. Its floor is dark but because the walls are very low, good conditions of illumination are required to observe it to best advantage. Just in Frigoris lies Archytas (2) (31 km) and its neighboring little sister Protagoras (22 km). Archytas with its high walls and sunken floor is a prominent sight but the shadow hides its internal triple peak system. Mons Piton sticks out brightly with its peak rising to well over 2,000 meters and a summit craterlet.

To the east of Plato is a fine example of a rift valley – Vallis Alpes (3) – which cuts right through the Montes Alpes from Mare Imbrium to Mare Frigoris. There is no other similar feature on the Moon. It is 166 km long and 20 km at its widest. It was formed through the sinking of the land mass between two cracks in the surface as the lunar crust was subjected to the stresses of settling following the impact responsible for the Mare Imbrium. This type of fault is geologically known as a graben. The Alpine mountains form steep walls at either side rising to an average of 2,000 meters.

Do inspect images for detail by excessively enlarging on the computer (figure 10.18). You might just be able to see (but not here) a central rill in the valley carved by the action of lava flowing from Mare Imbrium. Also at the bottom of the enlarged picture can be seen Cassini at 56 km diameter with irregular walls, one deep smaller crater within and a second giving the feature a very well recognizable pattern. Also to notice in this enlargement is the rut towards the northwest "corner" of Egede possibly gouged out by an incoming meteorite after the formation of Egede. Visibility of greater detail brings out nicely the relative heights of mountains through inspection of the lengths of shadows.

Some systems are grouped together and instantly recognizable, such as the crater chain Catharina (1) (104 km), Cyrillus (2) (98) and Theophilus (3) (110) that border Mare Nectaris (4) easily picked out by even the most modest of telescopes or binoculars (figures 10.19 and 10.20). Catharina, the oldest of the three, contains an interesting internal structure of a shallow ring with low ruined walls;

Figure 10.19.

Figure 10.20.

indeed it has five such craters within. There is no central peak. Both Cyrillus and Theophilus have mountain groups within. The walls of Cyrillus are somewhat better preserved and some terracing can be seen here. The interior of Cyrillus is a busy place with craters and peaks, one rising to over 2,000 meters. Its perimeter has been dented by the more regular but spectacular Theophilus with its terracing, flat floor and massive central peaks rising to 1,400 meters. Around Theophilus can be seen many ridges and ditches and in places small "ponds" of lava that formed from the descending ejecta while still molten.

The scarp of Rupes Altai (5) runs 427 km past this group to high-walled Piccolomini (6) (87 km). The Rupes is another of the majestic dominant features of the lunar landscape and rises to an average of around 2,000 meters in places and down to surface level in others. From this wide view the possibility can be seen of a much larger Mare Nectaris existing whose lava flow retreated leaving the Rupes scarp. Fracastorius (124 km) to the south of Nectaris is a bay formed from a huge and ancient impact crater that filled with lava and sank or suffered erosion to the northern wall. To the east of Theophilus lies the 28 km conspicuous bright crater Mädler guarding the entrance to Mare Tranquillitatis. Close by is the sunken or ghost crater of Daguerre (40 km). Once again, do computer magnify to enable inspection for interesting details (figure 10.21). It also enables counting of the inner craters of Catharina.

The following pictures show the same group and scarp illuminated from the opposite direction to illustrate that rarely is the same feature seen in exactly the same light twice (figures 10.22 and 10.23). The time of night, time of year, viewing conditions, and so on, all contribute to the excitement of seeing something new. The small bay to the north of Theophilus is riddled with rocks, ridges and ancient

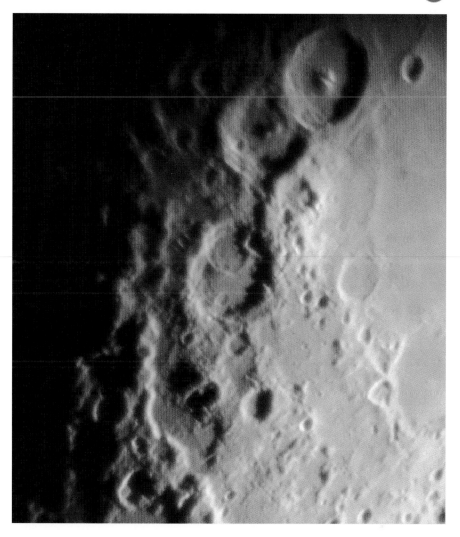

Figure 10.21.

marks, suggesting the flow of lava in early times. A very different view of the inner structure of Catharina and Cyrillus is presented. Wrinkles and ridges in Mare Nectaris and smaller surrounding craters also vary in appearance and detail.

Other groups are close by – Ptolemaeus (1) (164 km), Alphonsus (2) (108 km), and Arzachel (3) (96 km) with nearby Albategnius (4) (114 km) and Hipparchus (5) (138 km) (figures 10.24 and 10.25). Also Walter (6) (128 km), Regiomontanus (7) (129 km) and Purbach (8) (115 km) to the south. Features in this area are fre-

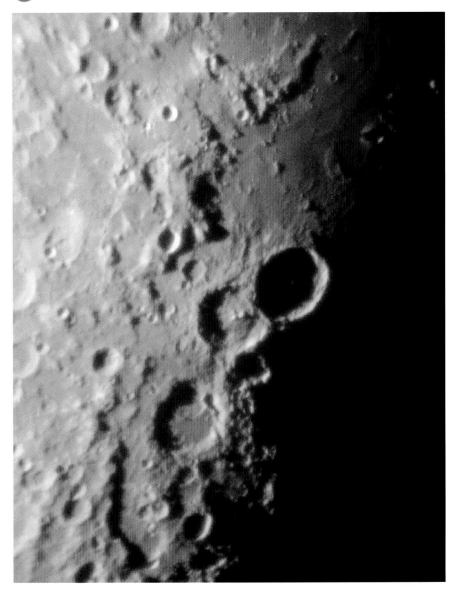

Figure 10.22.

quently observed and photographed not only for their beauty but also due to the convenient time of night they appear at the terminator.

As usual each crater has its own peculiarity or pattern. Ptolemaeus is huge with a relatively flat floor that allows the crater Ammonius at only 8 km to have pride of place. Its walls are fairly well preserved and rise to over 2,500 meters above the darkish floor. To the south-west lies a flooded plain with the conspicuous Davy (35 km) sitting on its south-western wall. On the south-eastern wall of Davy itself sits a tiny but bright and well-formed crater, Davy A. Alphonsus is a grand sight through the telescope with its wide and craggy walls. The floor reveals much

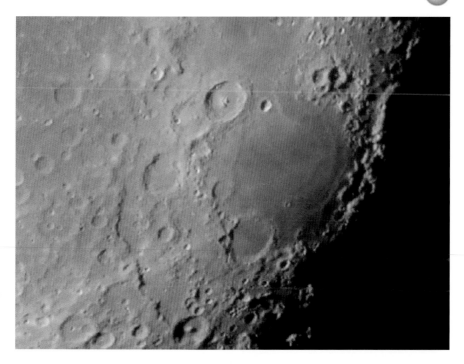

Figure 10.23.

Figure 10.24.

Figure 10.25.

structure, some of which can just be discerned in the image, of wrinkles and craterlets as well as rills and a modest central peak rising to 1,500 meters. Arzachel is another wonderful sight, its high, wide and terraced walls reaching to well over 4 km. Inside can be seen a prominent peak and a well-defined offset crater. A fascinating crater snuggles up between the southern wall of Alphonsus and the north-west wall of Arzachel. Alpetragius is symmetrical and only 40 km wide but a very deep 4 km. Most notably and worthy of close inspection is the unusual interior having no floor as such, just an enormous broad domed peak. By contrast the nearby Herschel, to the north of Ptolemaeus and almost the same size and depth, is an irregular crater with a flat floor and a central peak. Albategnius is full of structure and detail – a pleasing sight at most magnifications – with high walls rising to over 3,000 meters.

Once again an opportunity has arisen to capture an image with opposite illumination. In this image (figure 10.26), just to the north of Ptolemaeus, can be seen the relatively small (41 km) sharp crater of Herschel, mentioned above, that is very deep at 3,770 meters. Hipparchus is huge but not quite as big as its neighbor Ptolemaeus. The walls are heavily eroded, giving a decidedly craggy appearance. As the morning or evening Sun dips into the hollow the floor reveals a small crater set into the northern wall, a prominent central peak and the large half-buried circular structure. Horrocks at 30 km is easily seen.

To the east of Arzachel and Purbach lies Rupes Recta, The Straight Wall, although it is a fault line not a wall (figures 10.27 and 10.28). It is 134 km long, slopes gently (7 degrees) and is well-known due to its likeness to the shape of a sword. The western floor of the rift is only around 300 meters lower than the east. The "Sword" runs in a slight arc (not straight) from the Stag's Horn mountains in the south to a small craterlet in the north. The small crater adjacent to the Sword is Birt (3) (16 km). The picture was imaged using a higher magnification eyepiece and shows Deslandres (1) (256 km) below the Sword with its broken-down walls and Hell (2) (33 km) within. Note the dark or light Sword depending on the direction of illumination – the shadow or bright slope face.

Figure 10.26.

Here, is a glorious image of an area of the south-eastern Moon showing the Sword, major groups of craters and the overlaying rays from Tycho (figures 10.29 and 10.30). The sharp angle of illumination and good seeing conditions highlight the tips of several central peaks such as Ball at 41 km on the edge of Deslandres (1), Regiomontanus (2) and the inner dome of Alpetragius (3). The outline of Clavius (4) is just picked out in the darkness of the terminator.

Rills or Rimæ are narrow crack-like features in the surface of the Moon (figures 10.31 and 10.32). But they are not simple cracks. They are fault lines where a

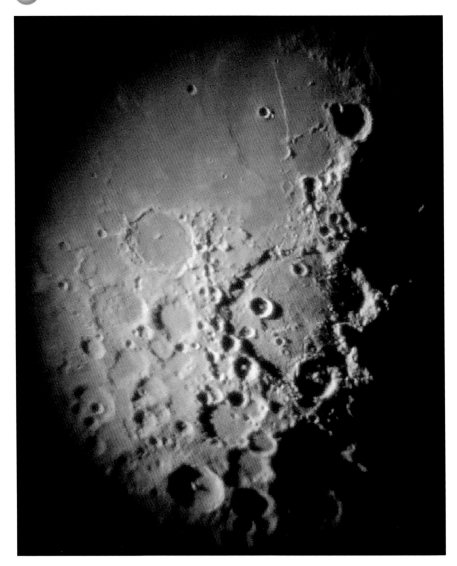

Figure 10.27.

central stretch of land has sunk leaving two vertical faces either side of a furrow or where early lava flows have worn away a long channel. Rills can also be composed of a line of adjoining small craters. The structure of one lunar rill was confirmed when the astronauts of Apollo 15 visited one such – Hadley. Rima Ariadaeus (1), discovered by Schröter in 1792, is a 220 km long rift up to 5 km wide and 2 km deep in places. As it was created it probably weakened the surface to allow subterranean venting and crater formation along its line. It is named after the small crater Ariadaeus (11 km) at its eastern end, picked out very

Figure 10.28.

brightly with its tiny companion in the evening Sun. Rima Hyginus (3) is also 220 km long and probably a fault line producing many small craters along its length. The advantage of being able to take so many pictures during one session is that later inspection of the images will reveal many features not actually noticed during photographing. Indeed, if the approach is to spot a feature and take a picture then much time and many pictures could be lost although clearly you might be after a particular target on some nights (figure 10.33).

Good, fast coverage taking a large number of photographs is a key part of success with the approach described. Thus the chances of catching a good seeing condition and making amazing personal discoveries (whatever you find is most likely to have been seen before by many) are maximized. Rima Ariadaeus (1) has been nicely captured here (figure 10.34) with the irregular shaped Julius Caesar (2) (90 km) just above and Rima Hyginus (3). Two craters show brightly at this illumination, Aggripa (4) (44 km) and Godin (5) (35 km) both with much internal structure and central peaks. Boscovich (6) at 46 km is easily spotted due to its dark floor, although its walls have been severely broken down. It is on the edge of the attractive small Mare Vaporum with its dark colored and varied surface relief. Within Mare Vaporum lies the brilliant Manilius (7) (39 km) with its terraced walls and central peak. To the south of Rima Ariadaeus a flattish area hosts a few brightly lit and very distinct craters – Cayley (14 km), Whewell (14 km) and De Morgan (10 km) with Dionysius (18 km) a little further on. Many broken or severely reduced craters and plains are apparent in the area, including

Figure 10.29.

Dembowski (8) (26 km) and the larger Lade (56 km) to the south of Godin whose southern wall has all but disappeared. Triesnecker (9) (26 km) is quite deep at 2,760 meters and is in an area of small rills and rimae. The two pictures, figures 10.34 and 10.35, exemplify different colors obtained at different times during the evening's viewing.

Figure 10.34 shows a nice mix of the Rima Hyginus, the double craters Murchison (57 km) and Pallas (46 km) and color contrasts within the area. The following image, figure 10.35, extends the view of this area towards the giant crater of Albetegnius at 137 km across.

Figure 10.30.

Figure 10.31.

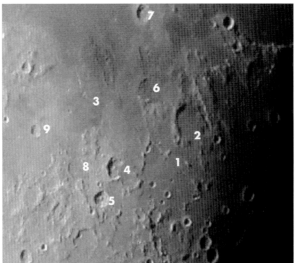

Figure 10.32.

No color filters, optical or computer, have been employed in any images shown. Varying colors such as a rich brown shown in figure 10.37 are obtained naturally from varying atmospheric and light conditions. Once again – be brave! Enlarge outrageously even to the extent of the picture becoming pixelated (figure 10.36). The tiny Rima Hyginus becomes a wide and jagged trench of small craters. Best of all, the even narrower rima of Triesnecker and its tributaries become apparent.

The Rima Hesiodus (figure 10.37) can be picked out stemming from the fascinating claw shape of Capuanus and runs for 300 km across Palis Epidemiarum to the "shores" of Mare Nubium.

Yet another set of scratch marks, Rima Hypatia, can be noticed from this wildly enlarged picture of a portion of Mare Tranquillitatis adjacent to the double craters Sabine and Ritter (figure 10.38). Just below the top crater (Ross) in the picture, Maclear, usually seen as a flat, dark ring, reveals its true crater form. The Arago domes are also very well picked out.

On occasions, while going through the night's pictures on the computer, an image pops up – you sit back in awe and amazement that you could have produced such an image (figure 10.39). Wow! The atmospheric conditions were right, the focus was good, and so on. OK – so Hubble can do better but we mustn't forget that we are talking cheap and cheerful. "Second-hand equipment, gross light pollution, no telescope temperature equilibration, standing on a heat sink patio, never done it before … and yet … look what I've got!"

The splendour of Copernicus with its layered crater walls, the bright rays of Kepler, tiny craters of 10 km or less and lovely coloring are all contained here. The picture is also capable of being enlarged to study in more detail. Definitely worth printing this one. The overenlargement now reveals more crater terracing and the fascinating "footsteps" in the surface between Copernicus on the left and Eratosthenes on the right (figure 10.40).

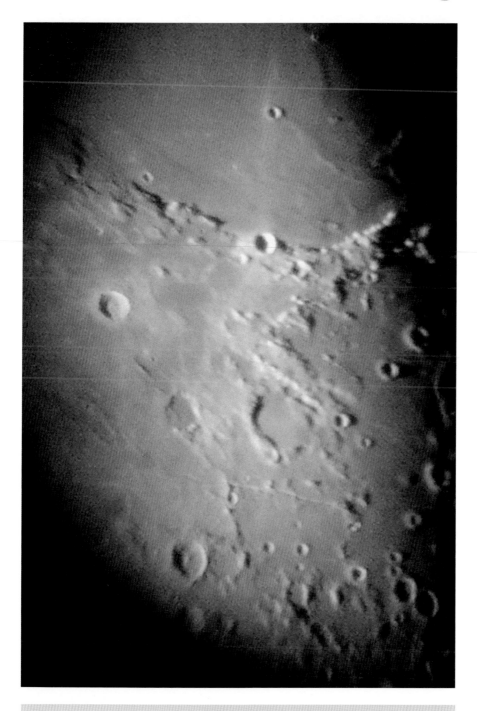

Figure 10.33.

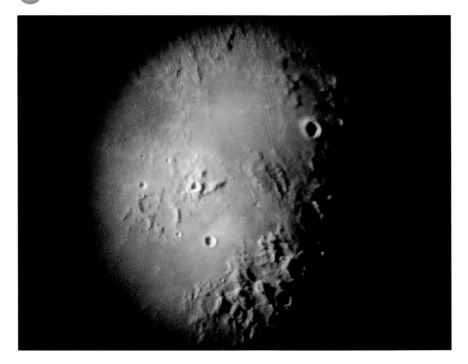

Figure 10.34.

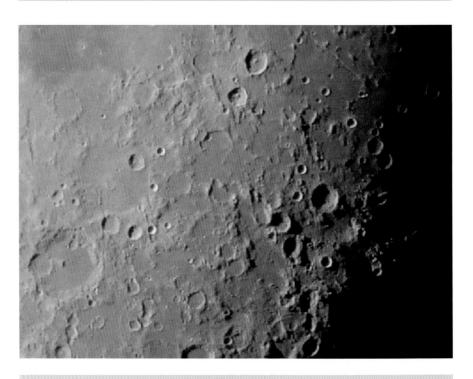

Figure 10.35.

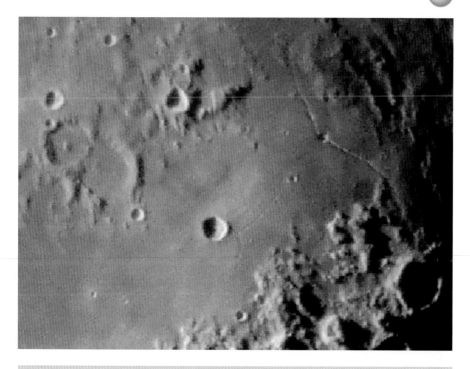

Figure 10.36.

It is difficult to praise Copernicus too highly and justly deserves one description as the Monarch of the Moon. It is not the largest crater at 93 km across but certainly has had a terrific impact on its surrounds. Most notably, even to the naked eye, it is the second largest origin of a ray system (Tycho being the first). The rays extend to many hundreds of kilometers across the area and the ray ejecta are tens of meters deep in places. This younger crater at a mere 900–1000 million years old is said to have come into existence following an impact from an asteroid up to 10 km across. Its walls rise to over 3,700 meters (up to 7,000 meters has been quoted!) from the floor and much terracing can be seen within. The size of the floor is significantly less than that between the wall tops since much in the way of boulders, rubble and debris has tumbled down in avalanches from the sides. The central mountains comprise three peak clumps, the highest rising to over 1,000 meters. The detail of the inner walls has been much described including a description of a cave-like formation to the north. Kepler, although much smaller than Copernicus, at just 32 km is the origin of the third-largest ray system and was formed at about the same time. Kepler has much less structure and detail and is somewhat shallower with walls rising to 2,300 meters above the floor. The light and dark radial spread from both craters fills the environment for hundreds of kilometers. However, some fingers of rays, emanating from Aristarchus many hundreds of kilometers away in Oceanus Procellarum, overlap those of the two dominant features Copernicus and Kepler. Clearly Aristarchus is much younger at around 300 million years old. The angle of illumination and the paucity of large

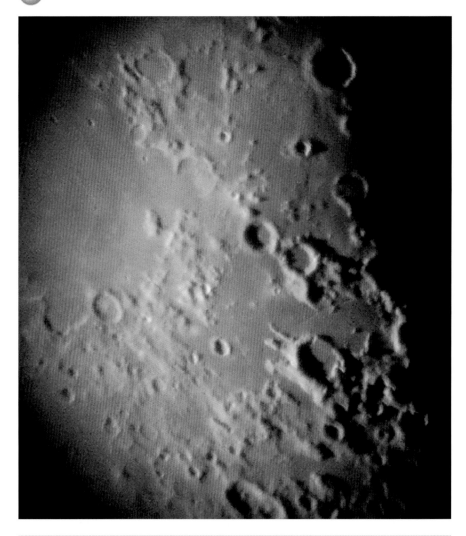

Figure 10.37.

craters in the area allow Fauth at only 12 km to stand out due to its shape being likened to a keyhole formed by association with a lesser crater. Not far from Copernicus to the east lies the barely visible "ghost" crater of Stadius at 69 km in an area peppered with craterlets, some forming a chain.

Copernicus at the terminator reveals its central peaks in shadow and its position among the maria of the area (figure 10.41), and in the following picture (figure 10.42) its radial ruggedness where the Montes Carpetus are also illuminated nicely. The diminutive crater Vinogradov at 11 km can be clearly seen as can other very much smaller craters and bumps exemplifying the resolution possible with the instrumentation.

Figure 10.38.

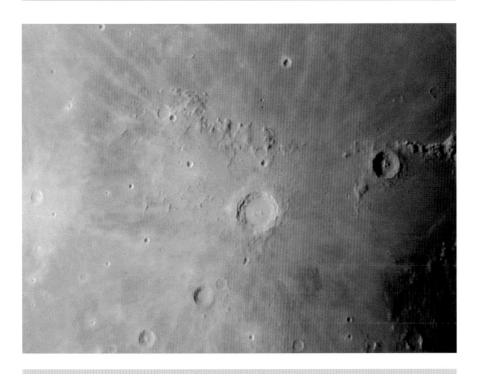

Figure 10.39.

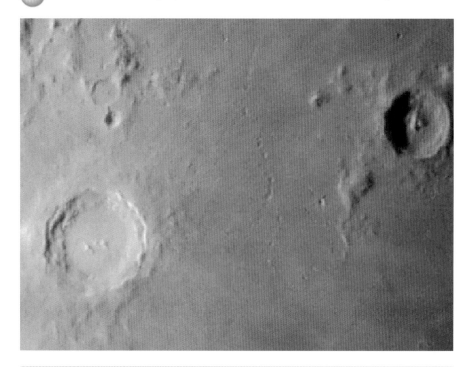

Figure 10.40.

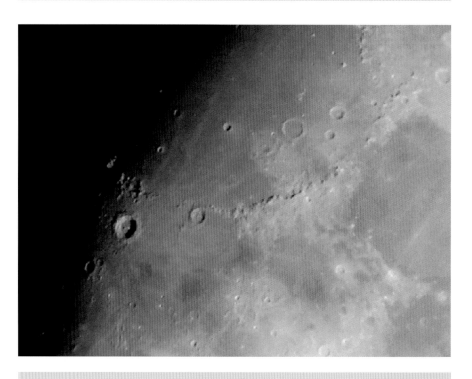

Figure 10.41.

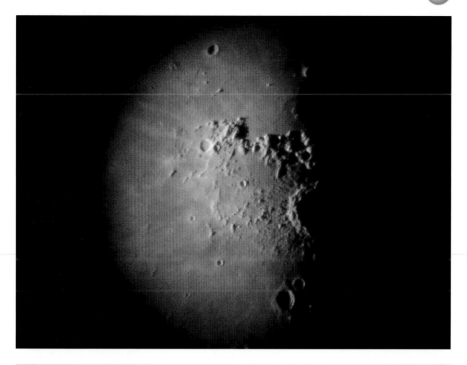

Figure 10.42.

This image of nearby Sinus Aestuum (figure 10.43) also competes for best picture not only for its clarity of detail and colors but also for picture composition (by accident!). The sunlight has just caught the Montes Apenninus beautifully. The rays from Copernicus, in addition to ripples and dark material, make the flat lava floor look very picturesque. It is well worth studying a lunar atlas to identify the many interesting features captured in this image. The 35 km crater Schröter (nowhere near Schröter valley) is badly broken down, while Mösting, the smaller bright crater (25 km) to the south, is well formed with a small central peak. Eratosthenes (58 km), the prominent crater to the north, is not dissimilar to Copernicus in its structure and beauty. The walls have some terracing and the floor has many small features. It is very deep at 3,570 meters. The plain of Sinus Aestuum has colorful impact ejecta and wrinkles and is striped with the more recently formed rays of Copernicus. Hard to imagine how one writer described the bay as "boring".

With a modicum of perseverance a complete atlas of the Moon can be created with your family digital camera and modest telescope but here are just a few more favorites.

Ripples of Mare Serenitatis (figure 10.44): Serenitatis is a 750 kilometer wide lava-filled impact basin. Two of its huge wrinkles have names. To the east lies Dorsa Smirnov, 130 km long and wending its way south from the narrow walled and feature-packed Posidonius to Montes Haemus in the south. Part of Dorsa Lister (290 km) curving round to Bessel can be seen to the west. Plinius, 43 km in diameter, stands prominently and brightly just inside Mare Tranquillitatis.

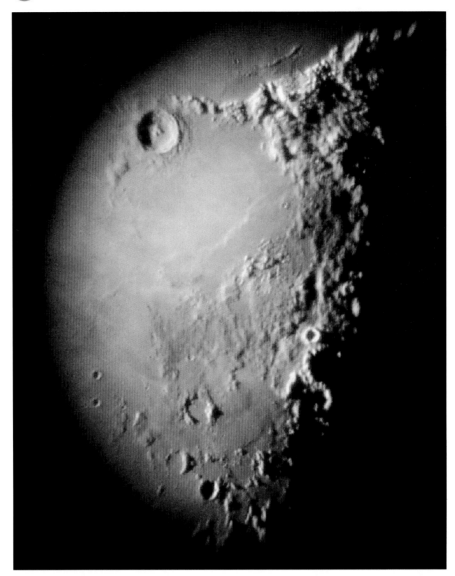

Figure 10.43.

Here is an unusually pictured Gauss and just about the whole crater has been captured at a very favorable libration (figure 10.45). The crater is big at 177 km across and circular, although the image shown is considerably foreshortened. Much detail of the complex floor is picked out by the sharp angle of illumination such as a central peak, ridge and crater. A pair of craters lies inward of Gauss. Hahn (84 km) has a very noticeable central peak and its neighbor Berosus is 74 km across.

Figure 10.44.

The wide view (figure 10.46), and a magnified portion (figure 10.47) of Mare Crisium highlights the rough and carved-out detail of the lunar surface with its myriad valleys, mountains, mini-bays, bright small rays and craterlets. Within the Mare many more wrinkle ridges, the result of contraction of the cooling lava, are evident. On its shores the faint Lick (34 km) and adjacent Greaves (14 km) lie close to the mountains that separate Mare Crisium from Mare Tranquillitatis. Picard at 23 km is also prominent in the Mare.

At a much more favorable libration (figure 10.48) the very deep and well-formed crater Neper at 137 km in diameter is sharply presented with its prominent central peak. Clearly visible to the north is Mare Marginis at 360 km across. An additional image (figure 10.49), at Full Moon, allows visibility of the Maria Smythii, Undarum and Marginis.

Also at a most advantageous libration (figure 10.50a) the 276 km diameter Mare Humboldtianum is clearly distinguishable beyond Endymion with a clear Gauss to the south.

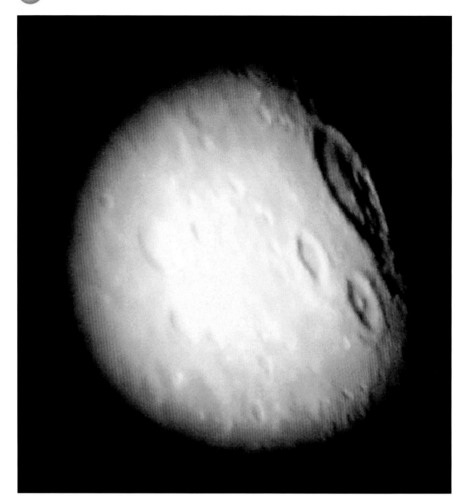

Figure 10.45.

The Full Moon, glorious as it is, conveys less crater information due to the paucity of shadows (figure 10.50b). Or does it? The mesmerism produced by the vision can easily drag the attention to brilliant rays to the exclusion of what might be found at the limb.

Just to the south of Mare Fecunditatis at the eastern edge sufficient shadow highlights the detail possible around the time of Full Moon (figures 10.50c and 10.50d).

Here the complete Humboldt crater (207 km) (not Mare) displays its inner lumpy structure. Photographs from manned flights around the Moon have revealed much complexity of rills and ridges on the floor of Humboldt invisible to Earth observers.

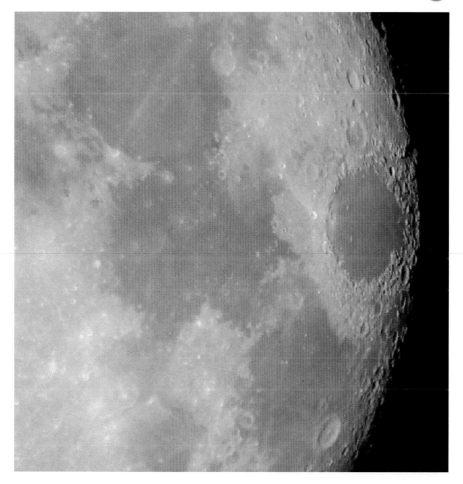

Figure 10.46.

Rugged features bordering Mare Nectaris are also shown in the image (figure 10.51). The joined craters of Isidorus (42 km) and Capella (49 km) have caught the light well. Isidorus has a flat floor with a deep small crater. Capella's large central peak is prominent and a crater valley intrudes right across. Many more small craters show themselves brightly including the normally obscure Leakey (12 km).

Lava flow colors, wrinkles and ridges, sunlit mountain tops and crater shadows all present themselves well in this image (figure 10.52), in addition to the Maria Frigoris, Imbrium, Serenitatis, Tranquillitatis and Vaporum and Sinus Aestuum.

The rich colors of lava flows in Maria Tranquillitatis and Serenitatis and the jagged appearance of the mountainous terminator are captured here (figure 10.53).

Figure 10.47.

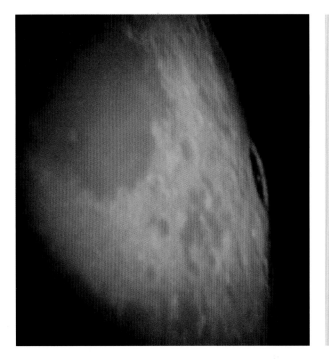

Figure 10.48.

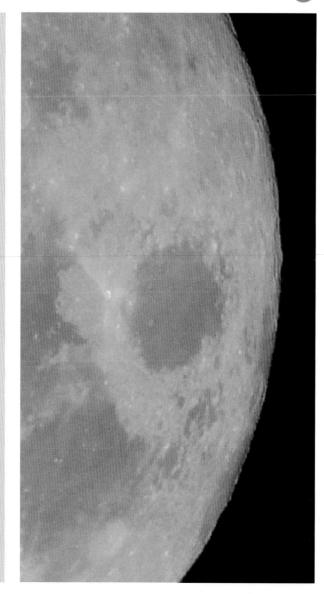

Figure 10.49.

Just dipping into the eastern edge of Mare Serenitatis are the Mons Argaeus (1), an area of massive mountains stretching for 50 km (figures 10.54 and 10.55). The picture shows sunlight at the terminator catching the tips of its many peaks, some rising to 2,000 meters high. Bright Vitruvius (2) at 29 km rests in the foothills but its walls are low. Following through Gardner (3) at 18 km lie a few buried craters just discernable in the Mare. A most satisfying prize caught at exactly the right conditions of illumination is a megadome 70 km across and unnamed, although referred to by Charles Wood as the Gardner Megadome due to its proximity to the crater of that name. Gleaming Proclus (4) (28 km), a source

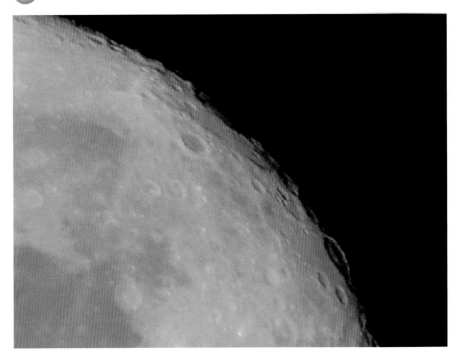

Figure 10.50a.

Figure 10.50b.

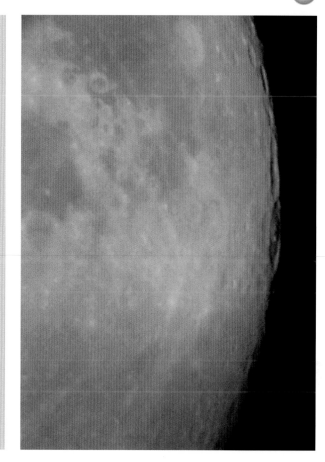

Figure 10.50c.

of a good ray system, just enters the picture. Rømer (5) (40 km) has high terraced walls to nearly 4,000 meters and contains a massive central peak.

Two pictures focusing on favorable librations are shown here. The first picture (figure 10.56) captures the twin craters of Cardanus (50 km) and Krafft (51 km) often responsible for the appearance of shafts of sunlight seen on a thin Moon just before Earth dawn. A particularly good libration is seen here. The outline of Seleucas (43 km), to the north of Krafft, is just visible with its high 3,000 meter walls although bathed in sunlight. From Seleucas, progressing towards the limb, is firstly Eddington at 125 km then the great expanse of Struve (170 km) combined with Russell (103 km). They are both vast flooded plains. The rim of Bartels at 55 km can definitely be identified beyond. Other craters and plains difficult to catch are just visible at this libration. Figure 10.57 displays the expansive crater Grimaldi (222 km) with its highly visible dark floor visible at the western limb even with the naked eye for those with good vision. The floor is not as smooth as it appears through low resolving power telescopes. Many rills, ridges and hills reside within and peaks rising to around 2,500 meters sit amongst the generally low and broken-down walls bordering this huge enclosure. As with Mare Crisium its shape is greatly foreshortened being at the limb, its east–west

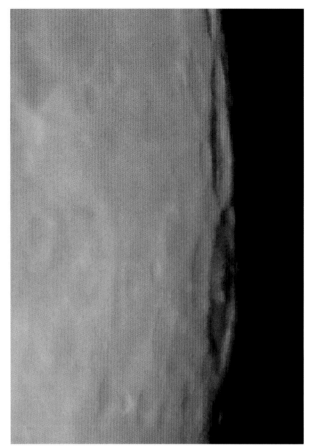

Figure 10.50d.

measurement actually being longer than its north–south. Riccioli, a little further out towards the limb and 146 km across, has a low wall and hills and rills inside. It is easily identified from the dark floor patch at its northern end. Although well illuminated, Hevelius (106 km) with its eroded walls and neighboring deep Cavalerius (64 km) can be seen to the north of Grimaldi. Beyond Hevelius towards the limb another great plain is in view, Hedin (143 km), with its heavily eroded walls. The Montes Cordillera are at the limb where Mare Orientale is to be found.

As mentioned, Grimaldi is a most prominent feature of the Moon to the west and generally appears as a big black unmistakable smudge. At such times it gives the impression that it is flat with little or no sizable walls or edges. However, when caught smack on the terminator its true characteristics are revealed (figure 10.58). The walls certainly are somewhat discontinuous but by no means non-existent with hills and mountains, the long shadows confirming their size. The blackness of the floor is apparent in addition to hints of ridges and craters soon to benefit from the Sun's rays. Hevelius (118 km) to the north via Lohrmann (31 km) is well walled and with much detail of craters, ridges and rills within. The convex nature of the floor is also perceptible in the picture.

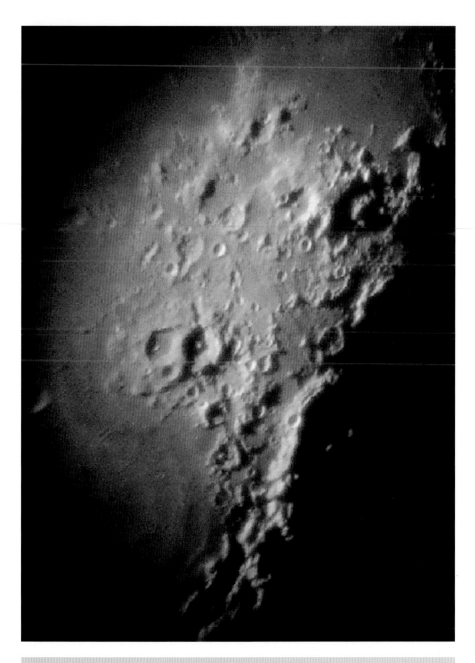

Figure 10.51.

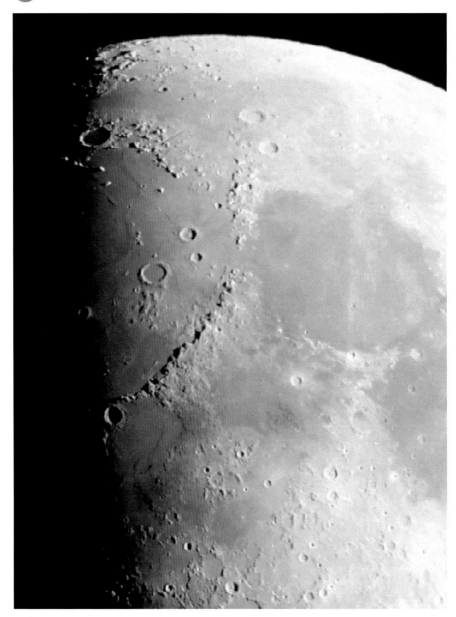

Figure 10.52.

Just to the south of Grimaldi lies a feature packed region (figure 10.59). A beautifully rugged plain sits right on the terminator, another reminder that very few areas are actually smooth but strewn with rocks and debris. Just as Grimaldi usually appears as a flat, dark patch so does Crüger in the rugged plain and Billy

Figure 10.53.

and Zupus (38 km) further in. Once again a keen eye and overenlargement (figure 10.60) is required to pick out a detail, a long rill, Sirsalis, which stretches for some 400 km beginning near the double crater Sirsalis. Its characteristics are possibly due to it having been carved from Mare Orientale ejecta during its formation. At its southern end it terminates with a strong arc to the east of the large crater Darwin (130 km).

Just one day away (figure 10.61) and the contrast between Billy and its neighbor Hansteen can be seen; plain Billy with its regular walls and dark flat floor and

Figure 10.54.

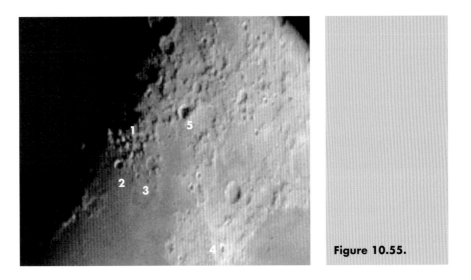

Figure 10.55.

Hansteen with its damaged walls but creased and bright floor. Zupus now shows as a small dark region rather than a crater but with high ramparts to the northeast. Rills and valleys abound in the area and Mersenius shows its convex floor.

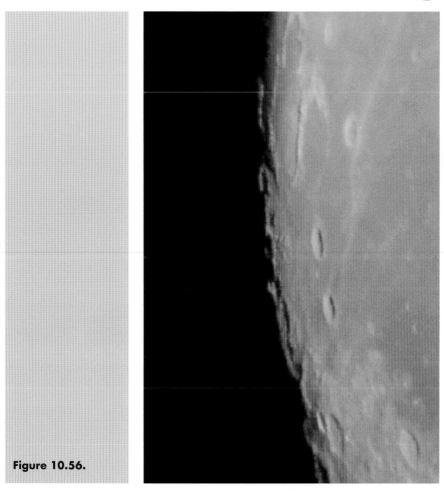

Figure 10.56.

Oceanus Procellarum has beauty in the simplicity of its wide expanse of lava flow punctuated by fascinating and prominent features (figure 10.62). Sitting seemingly on its own in the middle of a huge ocean is the sparkling arrangement of shapes and lights surrounding the Schröter valley (Vallis Schröteri). There is nothing like it anywhere else on the Moon. The features are mostly atop a large, smooth plateau 170 by 200 kilometers that is a little darker than, and 2,000 meters above, the surrounding Mare. Aristarchus is an extremely bright (the brightest) crater and the white spot can be seen at almost all times the surface is receiving the Sun (even frequently at Earthshine). Aristarchus (40 km) is a young crater probably just 300 million years old and its ejecta resulting from the forming asteroid impact is strewn far and wide as noted earlier. It has a small central peak. Herodotus at 34 km diameter is a much darker feature than its companion Aristarchus with a serious break in its northern wall. The star of the area is, of course, the valley that begins north of Herodotus at a 6 km crater and widens soon after at 10 km where it has the nickname of the Cobra's Head. It winds its

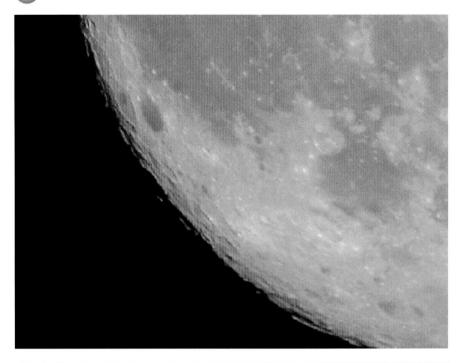

Figure 10.57.

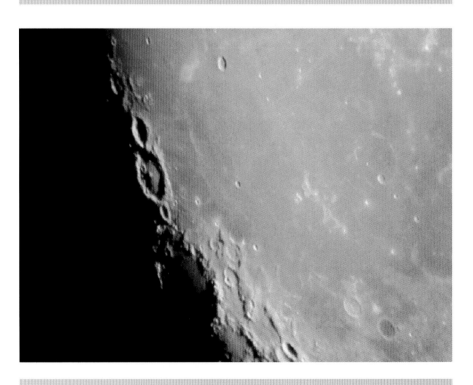

Figure 10.58.

Figure 10.59.

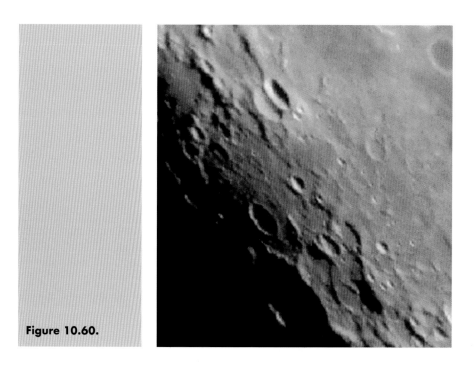

Figure 10.60.

Figure 10.61.

way for up to 200 km being 1,000 meters at its deepest and 5 km wide although reaching 10 km in places. The plateau is edged to the north by a thin ridge of mountains, Montes Agricola, running for 160 kilometers. The Montes Harbinger do not show too well in this image. They are essentially a group of hills rising to around 2,500 meters.

To the south of the valley lies the flooded crater Marius at 58 km with a low central hill. It is more noted for the area in which it sits. The largest field of domes on the Moon is to be found to the west of Marius nicely picked out here right on the terminator at a very favorable angle of illumination. One estimate quotes at least a hundred domes over a huge area. Tiny Bessarion at 10 km across shines brightly to the north of Kepler with a neighboring smaller crater. Many other small craters around 10 km in diameter can be seen in the relatively featureless expanse (figure 10.63). This enlargement of the Marius area provides visual evidence of the great number of domes and in particular that many have small craters in the center. The sharp feature of Reiner (30 km) lies to the south of the dome field on one of the many ridges in the area.

In pursuit of domes one must not be missed (figure 10.64). Mons Rümker is the largest lunar dome at 70 km across and can be seen reasonably well towards the terminator as a dark blob rather than a bright face as would be the case for a steeper mountain. It rises to around 700 meters and consists of several bumps in

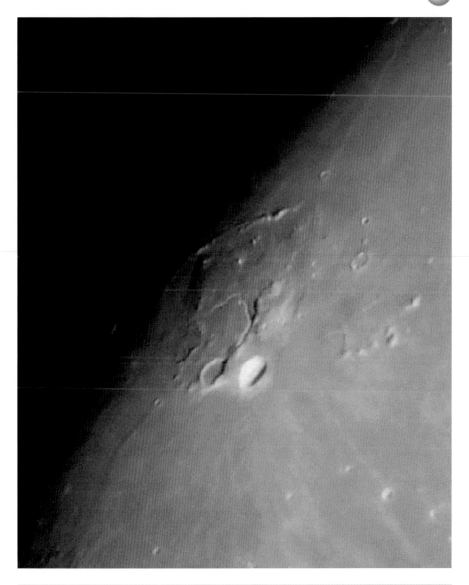

Figure 10.62.

one rather than a single dome. It has also been referred to as a semi-ruined plateau (and a buried tortoise! Sorry, that's mine). The enlargement shown in figures 10.65 and 10.66 reveals more of the gentle slopes.

The previous night's viewing presented Rümker bang on the terminator in such a way as to preclude any vision of detail. Perseverance might be rewarded with an improved view of this broad lump.

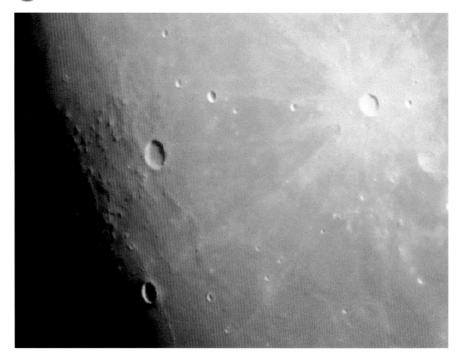

Figure 10.63.

Some remarkable features have been captured here amongst the crater-ridden south (figure 10.67). Schickard is a huge crater or walled plain and at 227 km across is one of the largest on the Moon. It has fairly well-defined walls rising to 2,500 meters in places although a breakthrough to the north-west has been caused by rubble and debris thrown out from an impact hundreds of kilometers away. The wide flat floor is also one of the most colorful, being much lighter in the middle than at either end. A few craterlets and hills are to be found within. Adjacent to Schickard is the well-known trio of Wargentin, Nasmyth and Phocylides. The light has caught the trio almost perfectly. Phocylides is large at 114 km with reasonably intact walls except at the join with the other members and a small crater neatly carved out of the south-east. Although the floor appears flat in this picture it does actually contain much detail and there may be a fault within. Nasmyth is the smallest of the three at 76 km. Illumination has been caught just at the right angle to reveal the unique feature of the lava filled crater Wargentin at 84 km. The lava has risen exactly to the rim at 300 meters above the surround to form a plateau that is relatively smooth, indicating its possibly younger age than the others. Actually, there are remnants of a wall at the rim, one segment rising to about 150 meters. Most angles of lighting fail to show the relief necessary to reveal its true structure.

Five well-formed craters line up neatly and prominently at this time towards the limb – Wilson (69 km), Kircher (73 km), Bettinus (71 km), Zucchius (64 km)

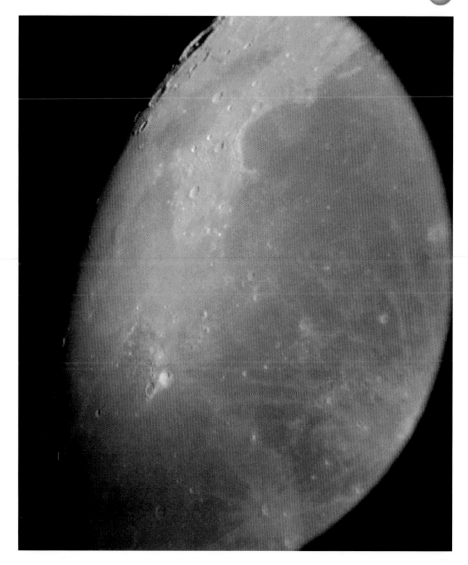

Figure 10.64.

and Segner (67 km). Bettinus and Zucchius have central peaks just visible here. These craters are dwarfed, as are most others, by the giant Bailly discernable beyond at the limb. At around 300 km in diameter it is by far the largest Earth-facing crater or walled plain. The walls are very irregular although some peaks rise to 5,000 meters. Its floor is littered with rocks, bumps and craters. This crater is nothing to do with Baily's Beads. They are named after Francis Baily (note the different spelling) who described the effect of sunlight shining through the valleys of the Moon at times of solar eclipse after the annular eclipse of 1836. "His crater"

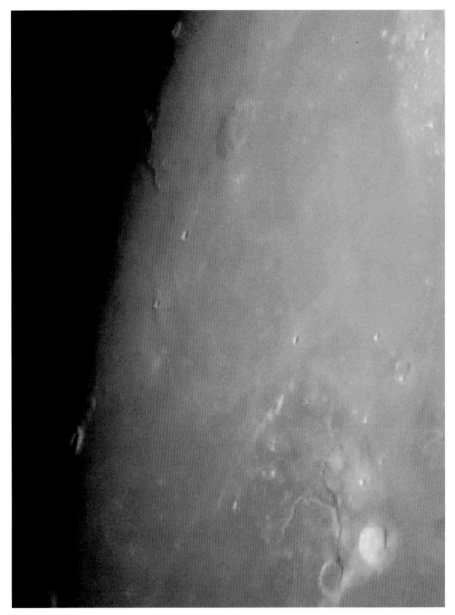

Figure 10.65.

lies in Mare Frigoris. Tycho can be seen spraying the area with its rays. Compare this image with that previously shown for Moon sector 12. Opposite illuminations highlight in relief a completely different set of features. Figure 10.67 presents Schickard and Wargentin to advantage with Bailly's wall shadow. The previously described Region 19 image, figure 9.28, highlights the Zucchius group and

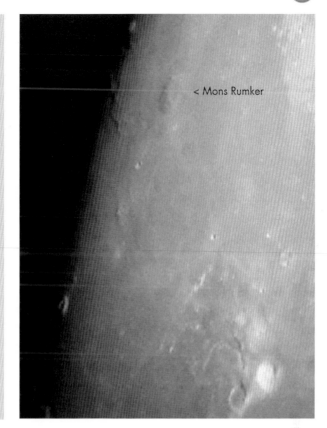

< Mons Rumker

Figure 10.66.

Figure 10.67.

Schiller and Hainzel in addition to revealing much more floor detail for Bailly even though having less shadow. A great libration for this area.

Schikard presents itself here right on the terminator (figure 10.68). Some of its internal structure can now be picked out and many small features also in the surrounding area. Ridges within the elongated Schiller are just illuminated.

The southern Moon image, figures 10.69 and 10.70, shows the wealth of craters and formations in the area. The eye is immediately drawn to the centerpiece, Tycho (1) and its radial streaks. Tycho is one of the deepest craters at 4,850 meters. Its diameter is 85 km and its walls are well formed and terraced. As crater sizes go Tycho is not special. Indeed if it were not for the spectacular rays splattered over such a large area it would hardly be noticed at all. At Full Moon its brightness causes complete domination to the eye. It has a central peak system rising to 1,600 meters above the floor. The glaringly bright visible rays overlaying everything else provide a clue to the young age of Tycho, the asteroid impact occurring possibly some 100 million years ago. It must be the youngest in the area.

To the west of Tycho three not so small bright craters impinge onto Heinsius (2) (64 km) breaking down its southern wall. Moving south from Heinsius appears sizable Wilhelm (3) at 106 km across. Being close to Tycho it is not surprising that on its floor can be seen ray ejecta striations. Its walls are wide and irregular but some peaks rise to almost 4,000 meters above the floor. It is difficult to see much detail under less than optimal light conditions. To the south of, and joining, Wilhelm are two irregular and eroded walled plains, Montanari at 76 km and Lagalla at 85 km separated by two craters the smaller being the brighter. The plains separate Wilhelm from a very large enclosure, Longomontanus (4) at 145 km. Continuing south, and coincidentally larger, the next significant crater is

Figure 10.68.

Figure 10.69.

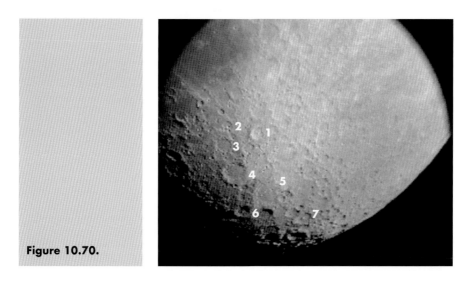

Figure 10.70.

the magnificent and photogenic Clavius (5). At 245 km across it is second only to the huge Bailly. Its mighty wide and eventful walls rise to almost 4,000 meters. Two reasonably-sized craters have contributed to the hard life of the walls. Porter at 52 km sits right on top of the northern wall while Rutherford (50 km) nestles on the inner southern terraces. Both have central mountains although those of Rutherford are more visible in this image. Most characteristic of Clavius is the curved line of distinct craters within, of gradually decreasing size. The four craters in order of decreasing size are: Clavius D (27 km), C (22 km), N (11 km) and J (10 km).

Two smaller (than Clavius), but by no means small, craters Blancanus (6) (105 km) and Scheiner (110 km) stand like twins further to the south-west. Both have high walls peaking at 4,000 to 5,000 meters and their dark floors are peppered with craterlets. To the south-east of Clavius lies the 114 km wide regular Moretus (7) with its prominent central peak.

Never the same image twice! The rugged vista of the southern Moon area constantly presents its multitude of craters in a different light. The grand southern Moon craters are seen here (figure 10.71) in opposite illumination and at higher magnification. The internal craters of Clavius are now much more clearly seen, especially those on the western wall. Over half of the 30 or so craters said to be visible with a 200 mm aperture telescope can be picked out. Previously almost invisible, the huge walled plain Maginus (in the center of the picture, figure 10.71, (194 km) now reveals its intricate detail and majesty, somewhat overawed by its proximity to the grand Clavius. The walls are eroded and the floor is pockmarked with small craters and debris. The array of craters large and small resting on the western wall form a superb mosaic in the evening Sun. At the top of the picture the corkscrew formation of craters is shadowed nicely – irregular Orontius

Figure 10.71.

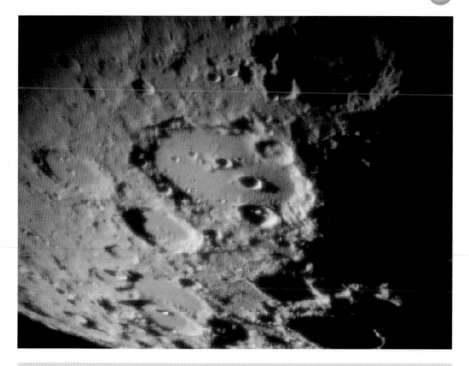

Figure 10.72.

(105 km), Huggins (65 km), Nasireddin (52 km) and Miller (61 km) with its nice central peak and busy floor. Just picked out right in the shadow of the terminator is the fascinating shape (often likened to a shoe) of Heraclitus at 90 km across with its irregular form and a central long ridge. It joins Licetus (75 km) and Cuvier (75 km). Stöfler at 126 km, with Faraday (70 km) sitting on its eastern wall, is another huge crater viewed to great advantage at the terminator. A nice shadow from the high western wall spreads across the dark and flat floor. Many more Clavius craterlets can be observed at even higher magnification in the enlarged image (figure 10.72). From a slightly different illumination most of them can be counted.

At a good libration for the southern Moon the huge complete crater Bailly can be captured as seen here south of the triplets Kircher, Bettinus and Zucchius, although not quite at the terminator to provide the clearest shadows (figure 10.73).

Many features are covered in this view of the area between Tycho (1), Gassendi (2) and Schiller (3) (figures 10.74 and 10.75). Western borders of Mare Humorum at the terminator show wrinkles and the characteristic Doppelmayer group (4) of reduced crater walls appear to the south. The enormity of Tycho's rays can be realized in this image, one wide ray stretching many hundreds of kilometers north past Bullialdus (5). Bullialdus at 61 km is an imposing crater and has been called one of the most perfect craters of its type. It has fine wide terraces on its walls and the floor is a very deep 3,510 meters below the rim, with a large group

Figure 10.73.

of central mountains. The characteristic shape of Hainzel (6) (70 km) with its combined craters begins to show itself near the terminator.

It can be a little difficult to photograph the narrow horns of the four day old (or less) Moon as it sinks towards the horizon in a darkening sky. The low angle of vision often experiences much atmospheric disturbance. The northern tip highlights beautifully the two prominent craters Atlas and Hercules (figure 10.76). The huge and complex 190 km Janssen area takes center stage with Vlacq (89 km) on this southern horn in figure 10.77.

The sunlit tip of Alphonsus' central peak is just one of the features sharply and artistically presented in the well-known trio of craters at the eastern edge of Mare Nubium (figure 10.78). To the south of Nubium (figure 10.79) lies the picturesque Capuanus (59 km), "claw", and Mercator–Campanus bordering the Palus Epidemiarum.

Take a closer look at some shadows. Jump in a helicopter and look down into the craters! Enjoy the experience of seeing the precise shape of majestic mountains etched across the crater floor. The 3D detail of what you see emphasizes the reality of what has become your own familiar backyard through the hundreds or thousands of pictures you will have taken.

In figure 10.80, Maurolycus outlines its jagged wall shape against a cratered floor. And under higher magnification the shape of a giant looms over the additional detail of craters and peaks (figure 10.81).

Figure 10.74.

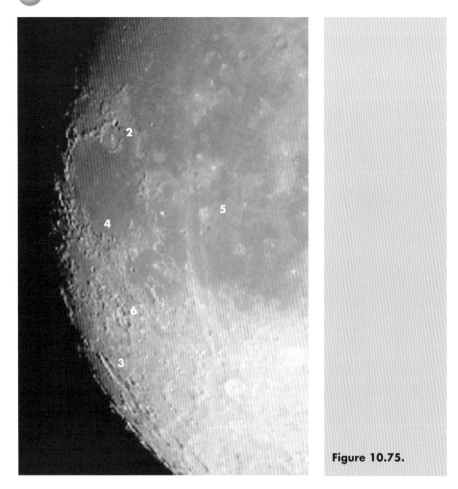

Figure 10.75.

Figure 10.76.

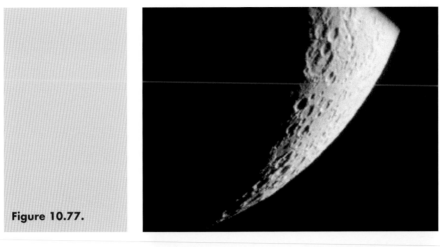

Figure 10.77.

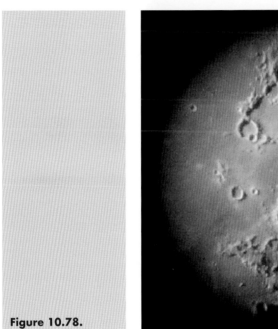

Figure 10.78.

Beautiful shadows can be seen resting on the floors of Aristoteles and Eudoxus bordering Mare Imbrium (figure 10.82).

The sharp pointed shadow of the central crater of Moretus reaches right across the basin (figure 10.83).

Archimedes' mountainous rim reveals its zig-zag form within, while mountains to the south cast a longer shadow (figure 10.84). Another sharp peak is shadowed here (figure 10.85) within Tycho, the origin of the dominating bright ray system, while the shadow of adjacent Maginus almost covers its floor.

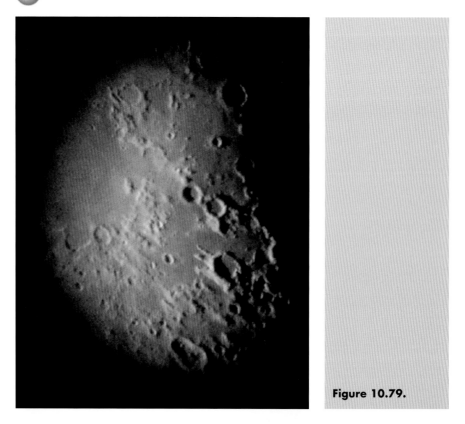

Figure 10.79.

Figure 10.80.

Figure 10.81.

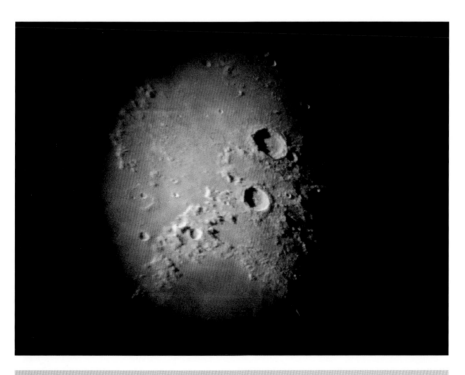

Figure 10.82.

Figure 10.83.

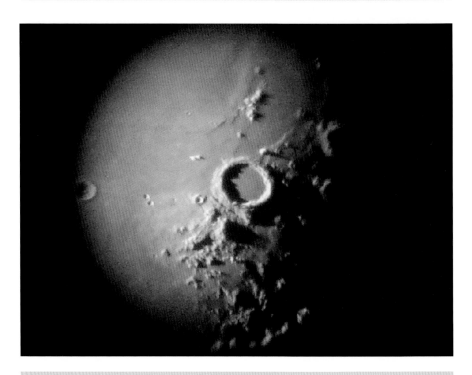

Figure 10.84.

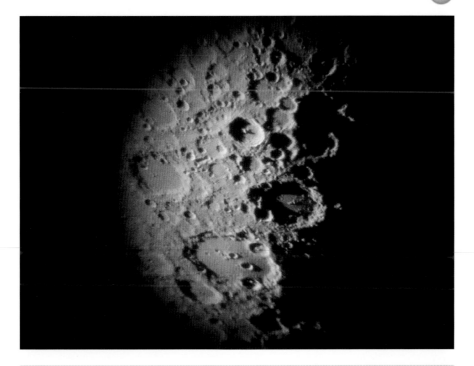

Figure 10.85.

The Sun brightens the rim and darkens the floor in this twilight image of Gassendi near the vast expanse of Oceanus Procellarum (figure 10.86). The opposite illumination in figure 10.87 reveals a rather flatter outline but highlights the 84 km diameter Mersenius with its distinctly convex floor.

Trivial, playful names abound for surface shapes resembling familiar forms. The "skull" and "baby" are to be found between the craters Diophantus, Delisle and Gruithuisen in Mare Imbrium. "The Helmet", close to Gassendi, is composed of material having different light-absorbing properties than the surrounding regolith as shown here (figure 10.88).

Also the shadows and brightness of the picturesque Montes Caucasus: just noticeable to the east of the break in the mountain ridge separating the Maria Serenitatis and Imbrium is the clear impression of the tiny crater Linné at only 2 km (figure 10.89). However, the light circular patch of ejecta seen is larger than the tiny crater within.

Wrinkles in the lava fields and flows present an amazing array of lines and colors when the illumination is just right and should be collector's items as for this image at the border of Mare Nubium next to the Sword (figure 10.90). Pitatus (97 km) with its central peak shows little shadow at this time.

Close inspection of this gross enlargement of the Sword area (figure 10.91) just reveals some intricate detail, only apparent during good seeing conditions, such as Rima Birt stretching from a small dome to the north and west, past crater Birt.

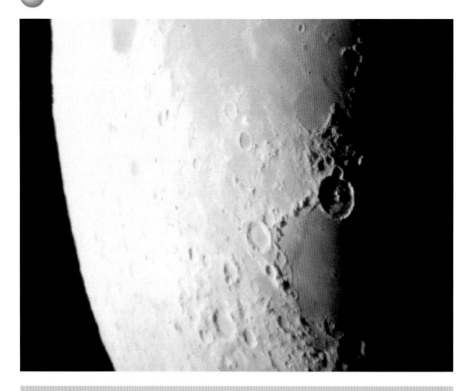

Figure 10.86.

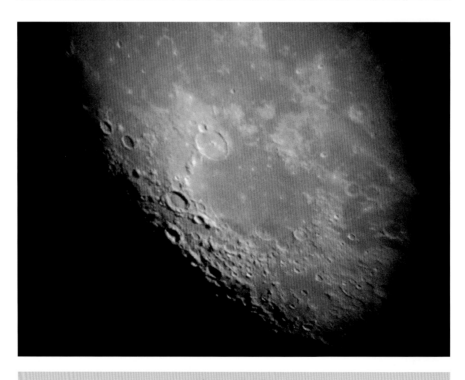

Figure 10.87.

Figure 10.88.

Figure 10.89.

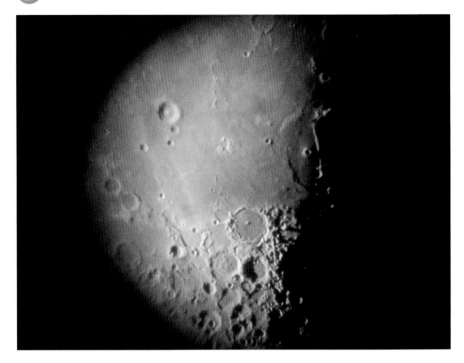

Figure 10.90.

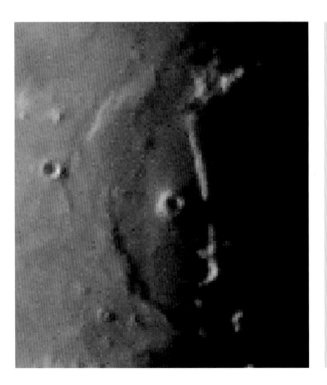

Figure 10.91.

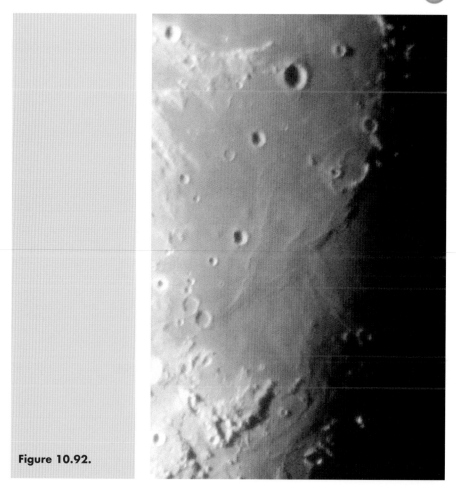

Figure 10.92.

More beautiful colors and wrinkles are visible towards the center of Tranquillitatis (figure 10.92). Carrel at 16 km across sits astride the walls of an old worn-down or buried crater. The complex nature of the ridges is clearly seen (figure 10.93). The picture is repeated here to enable comparison with an enlarged oppositely illuminated image of Arago where the most famous domes reside.

And, of course, the early and late phases of the Moon: capturing the last days of the old Moon does entail some dedication by the observer due to having to arise shortly before dawn! (figures 10.94 and 10.95).

How thin a Moon can *you* get? But do take extra care since the Moon is not far in your vision from the Sun (figure 10.96).

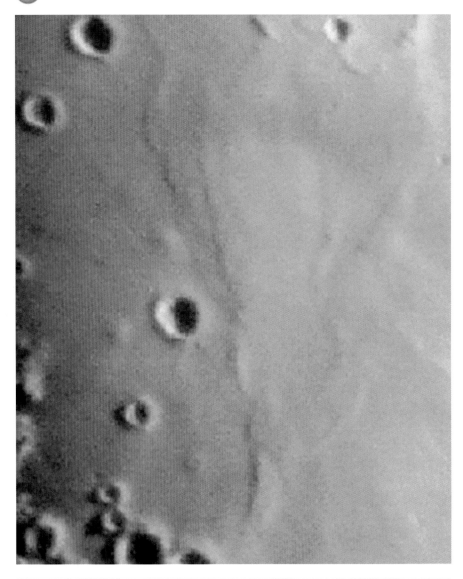

Figure 10.93.

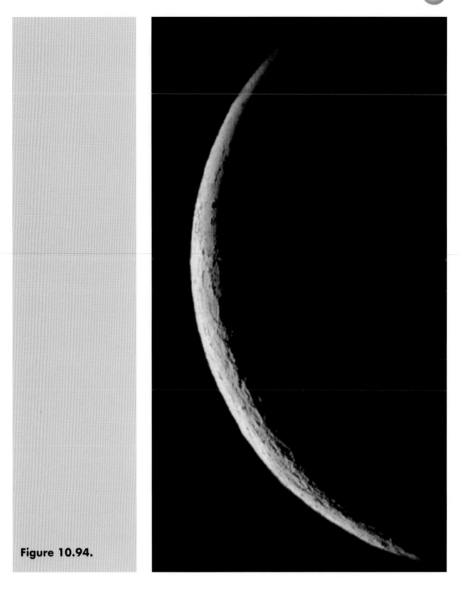

Figure 10.94.

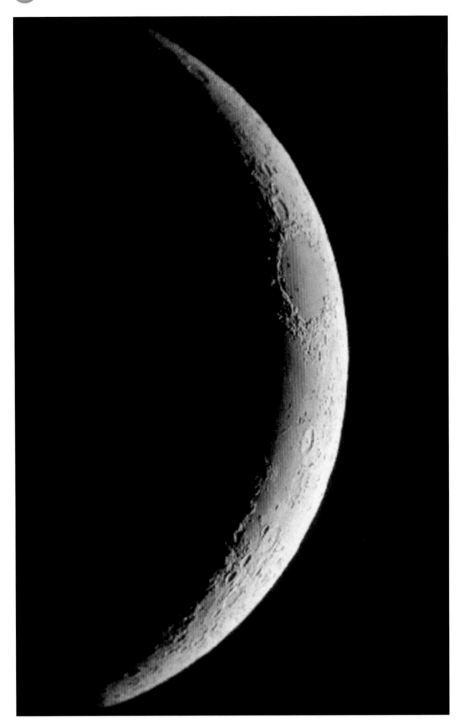

Figure 10.95.

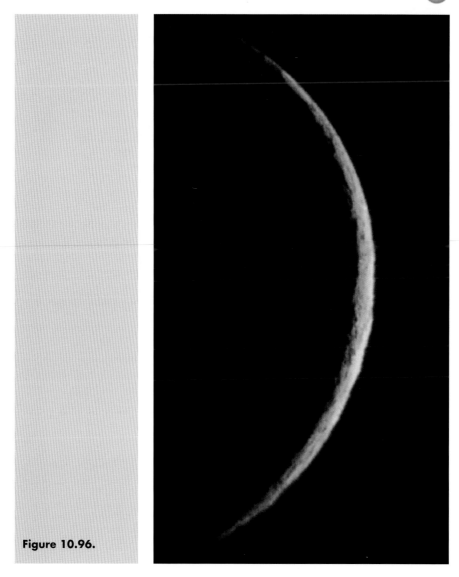

Figure 10.96.

CHAPTER ELEVEN

Lunar Events

1 Missions to the Moon

Introduction

The furious "space race" between America and the Soviet Union was given a kick-start when Russia successfully launched the first artificial satellite, Sputnik 1. The race to the Moon was on!

The early winner was clearly Russia in 1959 with Luna 1 passing within 5,000 km of the Moon and Luna 3 circling it and taking the first ever pictures of the far side. Zond were photographic missions the last of which, Zond 8, was launched and returned in October 1970. Twenty-four Luna probes were sent, the early ones achieving mixed successes. For instance, Luna 7 and 8 crashed but Luna 9 soft-landed and returned images. Luna 24, in August 1976, was the last of that era, successfully landing and returning to Earth with samples collected from the Moon's surface.

American attempts were heralded by the experimental missions of Pioneer and Able crafts from 1958 to 1960 all of which failed but they were experiments weren't they! Of the Ranger probes 1–9, the last three, 7–9, achieved their goal of a soft landing and the return of thousands of images. Of the 1–7 Surveyor probes from 1966 to 1968, only 2 and 4 failed to achieve the objective of returning many more thousands of pictures. Over approximately the same period Orbiters 1–4 returned hundreds of images and all except Orbiter 4 (uncontrolled and lost) underwent controlled impacts onto the surface at the completion of each mission. Four Apollo manned missions, 7–10, were flight tests around the Earth and Moon in 1968 and 1969 to prepare for future lunar exploration. Seven Apollo landings, 11–17, were planned but, of course, number 13 had to famously abort.

Many years later, in 1990, came the Japanese launch of Hagomoro to place the Hiten satellite into lunar orbit to return data until it crashed onto the Moon's surface in 1993. The American Clementine probe mapped a vast area of the Moon in 1994. Prospector, launched in 1998, searched for indications of the presence of water, particularly at the poles. Whether water does actually exist on the Moon, especially in useful quantities, is still under investigation.

Images of the Moon have also been returned to Earth from probes on their way to other Solar System destinations: Galileo to Jupiter (1990 and 1992), Cassini on to Saturn (1999) and NEAR–Shoemaker to the asteroid Eros (1998).

It is worth reminding ourselves of the enormity of the task and achievements to escape this Earth to travel nearly 400,000 km to another body in the Solar System; the technical marvels to build millions of components into a launch vehicle, a satellite, landing capsules and manned enclosures; the chemistry and physics to develop the fuel processes and mechanisms to safely propel the package into and through space. Above all, the brilliance and bravery of the dedicated astronauts themselves stand out in the field of human endeavor.

Now! Just where did all those probes land? The following pictures give a visual appreciation of approximate locations for successful (and some nearly successful!) arrivals. The landers are described in no particular order since some are necessarily grouped together in the same illustration. The opportunity has been taken to describe and discuss other features within each picture, which can be further identified and located from previous chapters.

Figure 11.1 gives some appreciation of just how many and how widespread were the landings.

Region 1

Luna 1, launched in January 1959, passed just a few thousand kilometers from the Moon, returning some information such as the almost negligible lunar magnetic field and studies of the solar wind. It then took up an orbit around the Sun possibly approaching within a few million miles of Mars but since communication from the satellite ceased 62 hours following its Moon bypass we will never really know.

Luna 2, launched in September of the same year, transmitted cosmic and solar data before plunging at speed into the Mare Imbrium followed moments later by parts of its propulsion system. The exact spot is said to be between the craters Archimedes and Autolycus. To commemorate this meritorious event the impact area was officially named Sinus Lunicus, the Bay of Lunik. The picture shows that Luna 2 punctured a most picturesque spot. Wrinkle ridges, valleys and mountains have all caught the morning light at the terminator. The crater wall shadow clearly reveals the central mountain peak complex in Aristillus. Much detail can be seen of the double crater Pallas and Murchison – Murchison's broken walls, Pallas' central peak and the small, deep adjacent Bode (figure 11.2).

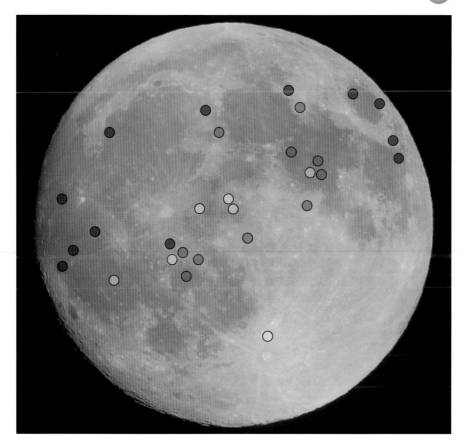

● Luna Missions.

◐ Apollo Missions.

○ Surveyor Missions.

◔ Ranger Missions.

Figure 11.1.

Apollo 15 was noted for its roving trips around the surface in a lunar "buggy". Many rock samples were collected, of interest because of their age and composition. David Scott and Jim Irwin had landed the Falcon at the foothills (well, within 30 km anyway) of Mount Hadley between the two Maria Imbrium and Serenitatis and close to the Hadley Rill. The pilot of the command module was Alfred Worden.

Figure 11.2.

Region 2

Apollo 11 arrived magnificently in July 1969. X marks the spot in Mare Tranquillitatis close to Moltke (6 km) (figure 11.3). The spot was chosen to be smooth and open to give the best chance for landing and communication. The first words as the module that descended from the command module Columbia touched down, "the Eagle has landed", those spoken by Neil Armstrong "one small step for a man, one giant leap for mankind" and Buzz Aldrin's "magnificent desolation" could never be forgotten. Michael Collins was the pilot of the command module. Some rocks were brought back (21kg in all) and the ALSEP (Apollo Lunar Surface Experimental Package) set up.

Surveyor 5 (September 1967) also landed in this area 25 km away, returning thousands of images and making specific analyses of the regolith (lunar soil).

Unfortunately the camera of Ranger 6, landing in January 1964, failed and no pictures were returned. Ranger 8, February 1965, was more successful, returning over 7,000 images before impacting into Tranquillitatis.

Some of the magnificence of the desolation screams out from this image. The beautiful lava wrinkles emanating from the ghost crater of Lamont, the gradient of rich colors from bright to dark and the sunlit hills beyond Moltke. The picture has been captured at a most opportune moment – right on the terminator. The low angle of illumination brings out clearly the sharp outline of the craters Arago,

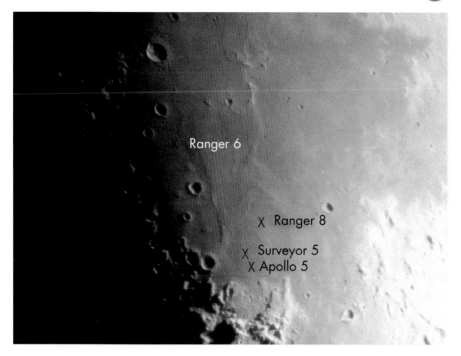

Ranger 6

X Ranger 8

X Surveyor 5
X Apollo 5

Figure 11.3.

Maclear, Ross and Plinius as well as the double twins around Sabine and Ritter, below which the Rimae Hypatia can be discerned. The photograph could not have been taken at a better time to pick out the difficult targets of the Arago domes, alpha and beta. Just to the east of Plinius many low-walled features are clearly visible: the low-walled Jansen (23 km), the tiny Cajal (9 km) and surrounding low hills and ridges.

Region 3

Apollo 12, command module Yankee Clipper and lunar module Intrepid, arrived in November 1969 in Oceanus Procellarum between Copernicus and the Montes Riphæus (figure 11.4). The crew of Charles Conrad, Alan Bean and Richard Gordon successfully achieved not only the recovery of more rock samples but also some parts that they snipped from Surveyor 3 for later analysis. They had set down only 200 meters from that previous probe. Surveyor 3 (April 1967), in addition to transmitting photographs, had determined that the composition and consistency of the surface would take the weight of the future Apollo 12.

The destination of the failed but thankfully "astronaut-safe" Apollo 13 was intended to be near the ruined crater Fra Mauro. Apollo 14 continued in that ambition and successfully landed there in February 1971. Stuart Roosa was the

Luna 5
Apollo 12
Surveyor 3
Apollo 14

Figure 11.4.

pilot of command module Kitty Hawk. Exploration of the surface after landing the module Antares was made more extensive through the employment of a lunar cart used by Alan Shepard and Edgar Mitchell.

The Russian Luna 5 crash-landed in May 1965.

Fra Mauro can just be picked out in the terminator lighting conditions as can its associated Guericke group of craters. The terraced walls of Copernicus and the rays of Kepler present themselves in this photograph.

Region 4

Apollo 16 landed in April 1972 between Dolland (11 km) and Descartes (48 km) (figure 11.5). The crew were John Young, Charles Duke and Thomas Mattingly, the lunar module Orion and the command module Casper. Once again a lunar rover was employed to gather surface rocks, of some significance to unravelling the mysteries of the Moon. Over 20 hours of roving was achieved.

A crater valley runs from the walls of Abulfeda (65 km), past Almanon (49 km) and on to join the Rupes Altai scarp. Tiny valleys can be seen running into the broken-walled Hipparchus from the environs of Halley (36 km) and Hind (29 km).

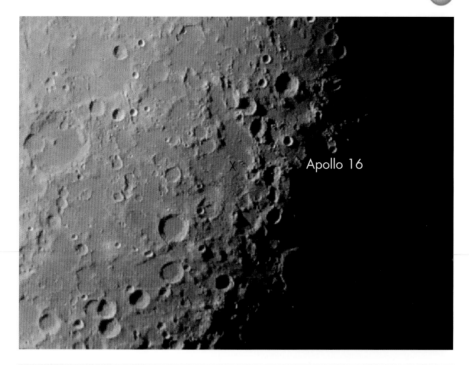

Apollo 16

Figure 11.5.

Region 5

Apollo 17, the last of the Apollo missions, landed in December 1972 between the Maria Tranquillitatis and Serenitatis in the area of Taurus–Littrow (figure 11.6). The command and lunar modules were America and Challenger and the crew Eugene Cernan, Harrison Schmitt and Ronald Evans. This essentially geological trip – Dr Harrison Schmitt was a qualified geologist – saw the lunar rover vehicle spend 22 hours collecting samples. The star of the show must surely be the discovery of a bright orange band around a small crater as they drove towards the South Massif. Samples of the soil turned out to consist of glassy beads of meteoritic impact origin. This picture has been somewhat overexposed to bring out the ripples and colors of the lava.

Region 6

Seven Surveyor craft were planned to accumulate further knowledge of the lunar surface, five of which were successful. Surveyor 1 landed in Mare Nubium close to the ghost crater Flamsteed in June 1966 (figure 11.7). During its transmission it returned over 11,000 images.

Figure 11.6.

Although soft landing in a relatively featureless spot of Nubium, not far away is a much more interesting expanse. Some inner detail of the easily recognized crater Gassendi and its tiny companion is apparent. At the terminator lies the fascinating Mersenius (84 km) with its bulbous floor and high and terraced walls. The smaller Cavendish (56 km) sits beyond with its 2,000 meter walls pockmarked with tiny craters. To the north of Mersenius at the terminator lies Billy at 46 km across one of a few craters noted for its very dark floor. Just to its north, Hansteen (45 km) gives a hint of some of the bumpy floor and terraced walls.

Region 7

The exact positions of Surveyors 2 and 4 are somewhat uncertain (figure 11.8). They failed and crashed approximately where indicated. Surveyor 6 definitely

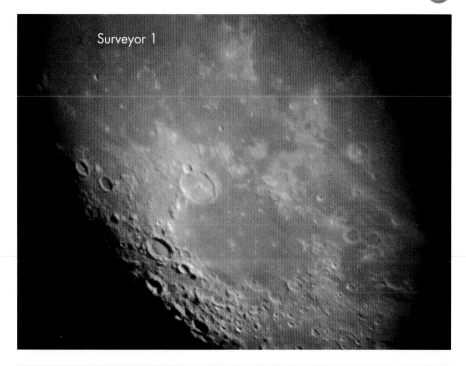

Surveyor 1

Figure 11.7.

landed successfully in the Sinus Medii and sent back almost 30,000 pictures in addition to information on the composition of the soil.

Even without enlargement the tiny craters of Blagg (5 km) and Bruce (7 km) can be seen in the center of Sinus Medii. To the north, bright Ukert (23 km) sits amongst a hilly terrain. Rima Hyginus is clear and Rima Triesnecker just visible towards the Mare Vaporum. The lovely dark hills within Vaporum itself are well picked out this near to the terminator. Much detail is highlighted within Albategnius and Apianus in addition to another take on the interesting Alpetragius with its domed floor. A lovely terminator here.

Region 8

Surveyor 7 landed in January 1968 among the highlands of Tycho returning not only over 20,000 pictures but much information on the chemical composition of the lunar soil, which confirmed the essentially basaltic nature of the regolith (figure 11.9).

Ranger 7 began its programmed self-destruct mission in July 1964, the objective being to take many high-resolution pictures before donating its body to the cause of lunar seismic information. It impacted into the area subsequently named

Figure 11.8.

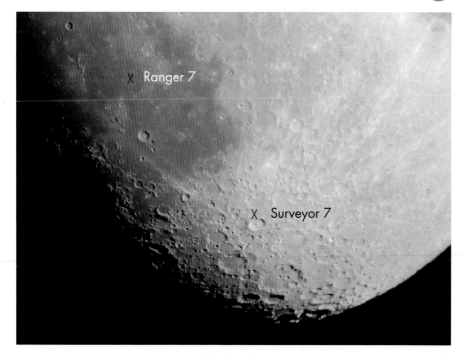

Figure 11.9.

Mare Cognitum to mark the place of arrival of this Ranger. Over 4,000 images were returned.

Tycho and the mountains of the south show clearly in this image. The stark difference is seen here between the crater Bullialdus (60 km) with its bright walls and interior detail and the adjacent almost invisible Lubiniezky (43 km) with its very low and discontinuous walls and a floor color close to that of the surrounding Mare.

Region 9

The last of the Rangers, number 9, arrived within the large (108 km) crater Alphonsus in March 1965 (figure 11.10). It was hugely successful, taking pictures containing detail not possible even with the largest terrestrial telescopes. The floor rills of Alphonsus, just visible in the picture, were shown to consist of a series of dents not cracks or valleys. Other valleys to the north-west of Alphonsus can be seen. Shadows bring into prominence the Rupes Recta and its associated crater Birt as well as many other features referred to previously.

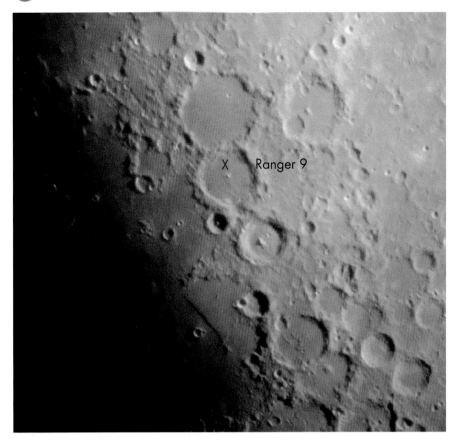

Figure 11.10.

Region 10

Lunas 7 and 8, both launched in 1965 were unsuccessful and crashed into the Oceanus Procellarum (figure 11.11). Also the two successful Lunas 9 and 13 both in 1966 landed in this area. The Luna 9 arrival area was subsequently named Planitia Descensus to mark the achievement. Many images and much data were returned that refuted a belief by one leading astronomer, Thomas Gold, that the lunar surface comprised a soft powder that would engulf any landing craft. Luna 9 did not sink but the reputation of Thomas Gold did! Luna 13, landing near Seleucus (43 km), made a more comprehensive photographic sweep of the area. The density of the soil was also tested and found to be not dissimilar to that of the Earth.

Seleucus can be seen brightly in the photograph and the smaller dark (in this image) Schiaparelli (24 km) on the way to Schröter's Valley. The dark-floored huge crater of Grimaldi can be seen, as always, under most conditions of illumination.

Figure 11.11.

Region 11

Five Lunas landed in the area of Mare Crisium, two of which (15 in 1969 and 18 in 1971) were unsuccessful. Lunas 16 (1970), 20 (1972) and 24 (1976) all collected soil samples and returned with them safely to Earth.

Figure 11.12 accentuates the clear-cut and totally isolated (from other mare) Mare Crisium with its rugged walls and dark interior. Several craters at the terminator are highlighted and have been described previously.

Figure 11.12.

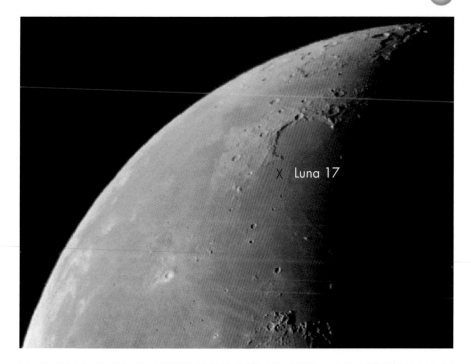

Figure 11.13.

Region 12

The Luna 17 mission (1970) was to transport a Lunokhod rover to the Moon to enable a greater range of investigation. It landed successfully in the Mare Imbrium near to the Promontorium Heraclides edging Sinus Iridum. For 322 days the rover progressed for over 10 kilometers southwards and back to the landing craft, taking thousands of pictures and many panoramic views.

The wide-scale view shown here (figure 11.13). allows appreciation of the relative positions of the Sinus Iridum, Oceanus Procellarum with the Schröter valley, Maria Imbrium and Frigoris and the beginnings of Montes Carpetus close to Copernicus.

2 Lunar Eclipse

It is easy to be discouraged by less than perfect weather. Providing the sky is not completely and constantly covered, an effort might be rewarded simply because you have recorded an event. Never mind that the quality is not up to expectations. An eclipse is a good example.

As for any object being illuminated, Earth casts a shadow into space on the opposite side from the illuminating Sun (figure 11.14). The Moon's orbit round

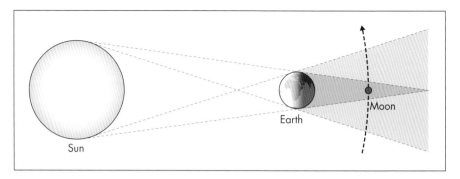

Figure 11.14.

the Earth, although complex, will regularly send it through the shadow either partially or totally. The diameter of Earth's shadow decreases with increasing distance as it points into space but is still very much wider than the Moon at the Moon's distance from the Earth. Eclipse totality can therefore last for an hour or more. At totality the color of the Moon can be from a beautiful bright copper with a clear atmosphere to almost black for heavily contaminated air following a significant volcanic eruption such as occurred in 1963 and 1964 following activity from Mount Agung on Bali and 1982 from the Mexican volcano El Chichon. Eclipses always take place when the Moon is full, of course. They can be seen from anywhere on Earth if the Moon happens to be in the sky at the time, which means that, over time, far more lunar eclipses than solar eclipses can be seen from any particular location.

The following images were obtained on the 9 November 2003, despite the efforts of clouds to spoil the night's viewing. (Figures 11.15 to 11.31)

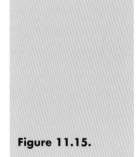

Figure 11.15.

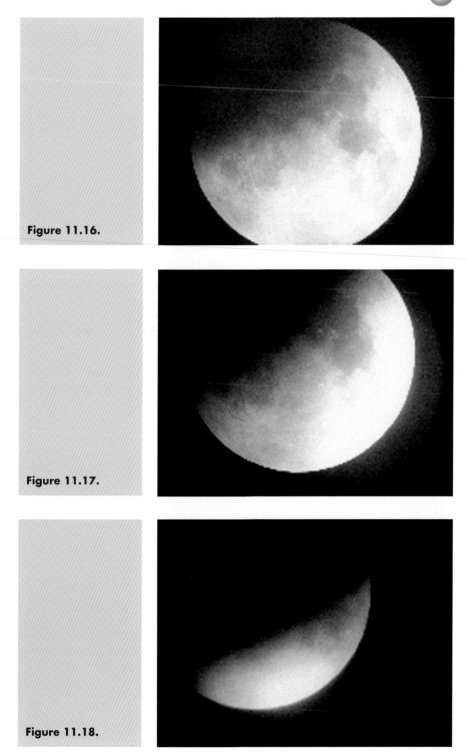

Figure 11.16.

Figure 11.17.

Figure 11.18.

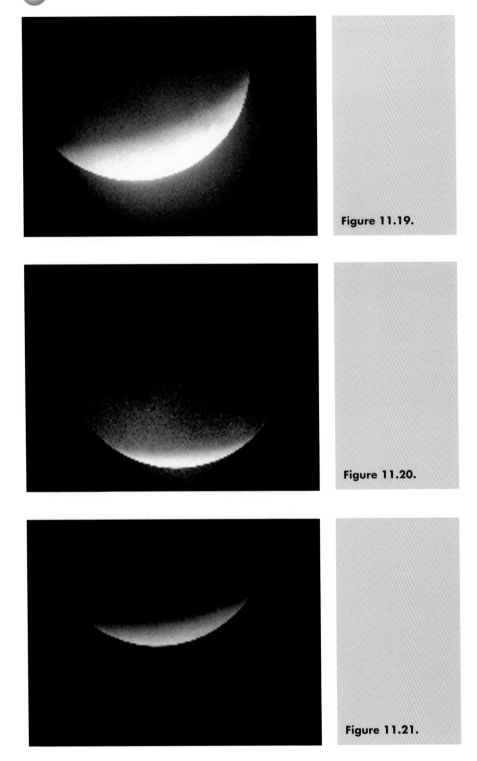

Figure 11.19.

Figure 11.20.

Figure 11.21.

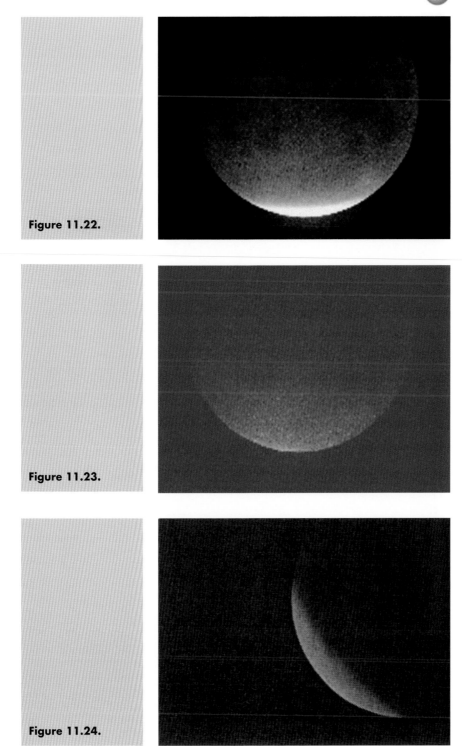

Figure 11.22.

Figure 11.23.

Figure 11.24.

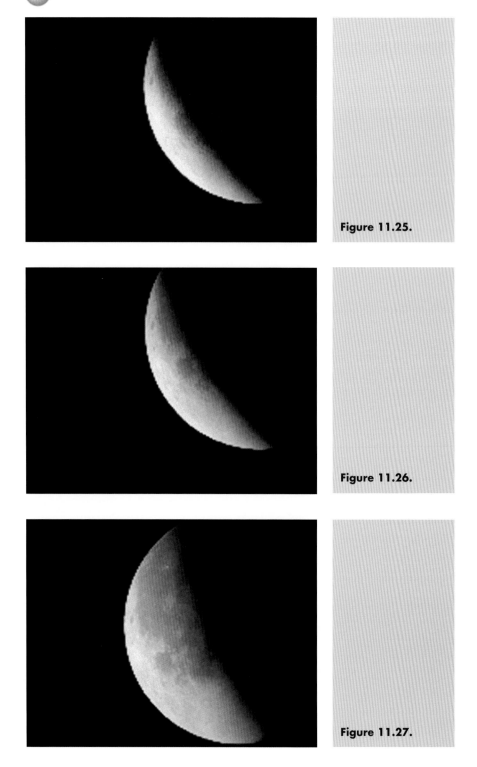

Figure 11.25.

Figure 11.26.

Figure 11.27.

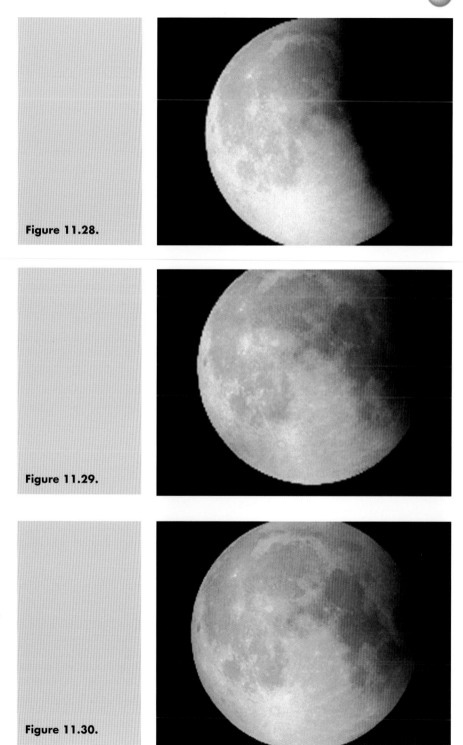

Figure 11.28.

Figure 11.29.

Figure 11.30.

Figure 11.31.

3 Earthshine

This picture (figure 11.32) was taken with the camera without the telescope very early in the morning. Jupiter can just be seen as a tiny white spot, bottom right. However, the naked eye could see the reflection of Earth's light on the remainder of the Moon – Earthshine.

Adding the camera to the telescope revealed little more …(figure 11.33)

… until the picture was brightened using the computer (figure 11.34).

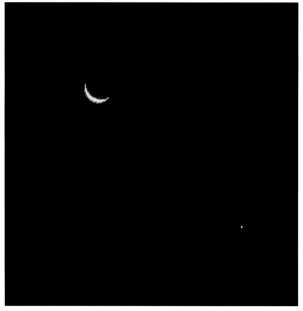

Figure 11.32.

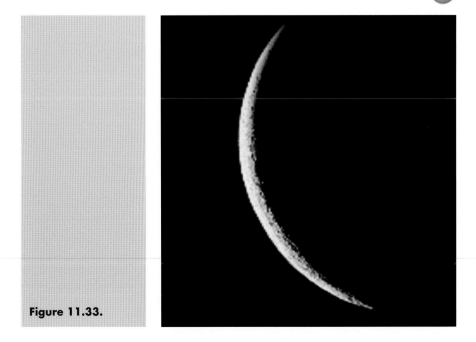

Figure 11.33.

Figure 11.34.

Figure 11.35.

One or two of the many pictures taken showed a lovely beam of light shining through the two craters Cardanus (49 km) and Krafft (51 km) (figure 11.35). In April 2005 the atmosphere was particularly clear and a terrific New Moon Earthshine was photographed (figure 11.36). So many major features were visible; the maria, Tycho's and other rays, Grimaldi, Aristarchus (of course, being the brightest crater), Plato and much more.

Once again the lesson is to take many pictures and watch for the unexpected.

4 Transient Lunar Phenomena (TLPs)

This picture does not look great. In fact it is nowhere near the quality expected. Bad seeing. (figure 11.37) But! Now enlarge it outrageously in the area of J. Herschel and optical effects can be seen (figure 11.38): red inside Robinson (24 km) and other spectrum colors at the terminator. Green hues catching the tops of high points over a large area are often seen. Clearly not in the league of TLPs, such as reported puffs of smoke; nevertheless, it is recorded, not just anecdotal, and a very firm lesson to study images closely as they are accumulated.

Figure 11.36.

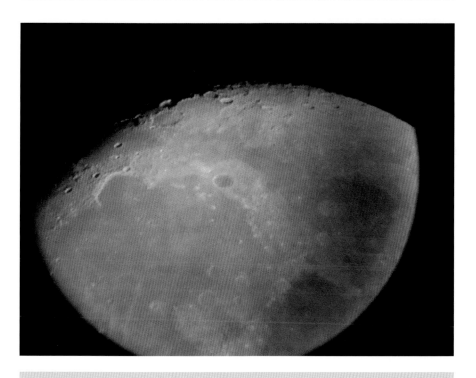

Figure 11.37.

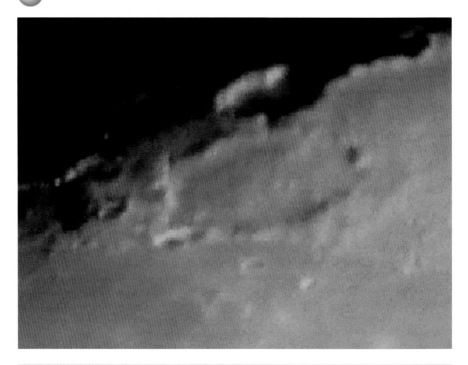

Figure 11.38.

5 A Chance Sighting?

As so often happens, careful and practised inspection of photographs taken reveal something unexpected and exciting. In the following picture a black spot was unnoticed at first glance against an uninspiring Moon landscape (figure 11.39). A possible speck of dust, if noticed at all, during anticipation of dramatic pictures tucked away within a night's work of a hundred or more snaps. Almost a year elapsed before revisiting the file. Questions began to form. Why is the spot so symmetrical? Why did it not occur in the picture before and the one after? Clearly it is unlikely to be a suddenly arrived and disappeared bit of dirt, it must be a photograph of something (figure 11.40). Why so black and in focus? A weather balloon? A child's helium balloon? A bird is a common cause. I discounted the possibility of an asteroid hurtling straight towards me for obvious reasons!

The chances might be small but it could be an artificial satellite. It certainly appeared in an area of sky where satellites frequently pass the field of vision during a night's viewing. What could be done to investigate? A simple calculation would eliminate the possibility if the size did not come anywhere near that of known satellites.

The calculation requires some known measurements to solve for unknowns in a triangle. Approximations must also be made. The distance of the Moon is on

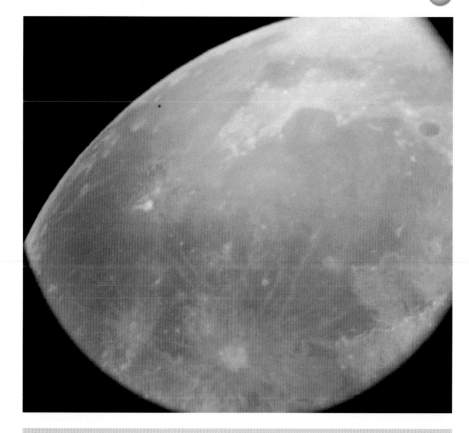

Figure 11.39.

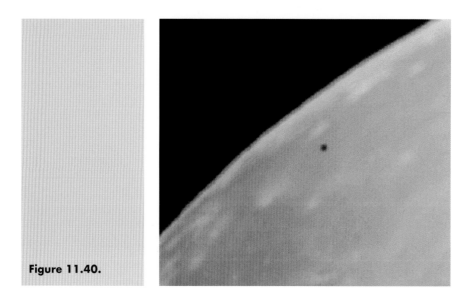

Figure 11.40.

average, from Earth surface to Moon surface, a little under 400,000 kilometers. It is not an unreasonable assumption that the hypotenuse adjacent to such a small angle in a triangle is the same length. Assuming initially that the spot was actually on the Moon (but clearly not!) its diameter could be estimated by comparison with craters of similar size. Of course a crater to be used must be in a central area to avoid errors due to foreshortening. In the image a likely target is Milichus towards Copernicus at 12 km across. The spot is a little smaller, therefore a reasonable estimate is 10 km, i.e. a radius of 5 km. The angle a is therefore:

$$\sin a = \frac{5}{400,000}$$

If we assume a typical distance for a satellite of around 800 kilometers (e.g. Envisat) the radius of the spot then becomes:

$$R = \frac{5 \times 800}{400,000} = \frac{1}{100}$$

A hundredth of a kilometer or 10 meters puts it well within the ball-park for a satellite. The dimensions of Envisat are $7 \times 10 \times 25$ meters.

An additional calculation of the speed of a satellite indicates that at the rate of 100 minutes per revolution around the Earth the spot would be a blur at 1/60 of a second camera shutter speed. Therefore it must be slow moving relative to such satellites.

Figure 11.41.

Attempts to identify an appropriate satellite passing at that exact spot on 14 August 2003 around midnight revealed possibilities but not certainties, but great fun anyway.

6 And Finally for the Moon ...

… the magnificent display of the almost Full Moon in all its glory showing the patchwork of maria and bright rays (figure 11.41).

CHAPTER TWELVE

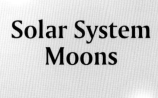

Solar System
Moons

Overview

The early-forming Solar System must have been a very untidy and busy place with all those bits of rock and dust bumping into each other to form the Sun and the planets. But even then there was so much more debris floating around often in chaotic orbits and trajectories. It had to go somewhere and much of it aggregated into lumps to form bodies not designated to be planets but captured by them. Hence the moons of our system and, of course, the asteroid belt. New ones are being discovered all the time as astronomical equipment and techniques advance. One current count puts the number at 136 although the definition of a moon must be constantly questioned.

Only two of our planets, Mercury and Venus, are without their own natural satellites. Both are too close to the Sun to have captured and retained moons for any length of time – too close to the planets and they would have been destroyed by gravitational forces, too far away and the Sun's gravity would have greedily consumed them.

Mars (Roman name for the god of war) is somewhat more lucky in having two moons, Phobos (panic) and Deimos (fear), named after the Greek mythological twins, the issue of Ares (Greek name for the god of war) and Aphrodite the goddess of love. Phobos flies closer to its planet than any other moon in the Solar System, just 6,000 km from the surface. It is, in fact, so close that it is gradually being drawn in by Mars and will eventually crash down or break into tiny pieces to form a ring in about 50 million years time. Martian inhabitants please note! Deimos, at $15 \times 12 \times 10$ km is the smallest of the major moons of the Solar System. Both satellites are thought to be small rocky asteroids captured with the assistance of Jupiter's massive gravitational pull. Both also have very low reflectivities (albedo) and, since they are so close to the planet itself, are difficult to spot.

Jupiter has a plethora of moons, four of which were first seen by Galileo as early as 1610 and are known as the Galileans – Io, Europa, Ganymede and Callisto. Every major moon seems to be accompanied by a superlative. Io is a hellish place, the most volcanically active moon in the Solar System, having around 300 active and inactive volcanoes in all. The volcanoes vent out sulfurous gases up to 300 km above the surface and at a speed of 1 km/sec. In contrast Europa's surface is smooth and white and covered in water ice and snow. It is therefore highly reflective. The streaks and lines criss-crossing the surface are likely to be caused by cracks in the thin surface crust with maybe an ocean of some sort beneath. Callisto is the most heavily cratered part-ice, part-rock satellite with its crust estimated to be 4 billion years old. There may also be an ocean deep beneath its crust. Ganymede is the largest satellite in the Solar System. Indeed if it revolved around the Sun instead of Jupiter it would be classed as a planet – somewhat bigger than Mercury. It is composed of a mixture of rock and ice and the surface has a variety of features such as craters in the darker regions and long furrows in the lighter ones. At last count there were 28 satellites moving round Jupiter and more are to be expected as technology and techniques improve the search.

Beautiful Saturn is also blessed with many moons – around 20 at last count. Titan is Saturn's largest, only a little smaller than Jupiter's Ganymede, with a thick atmosphere of nitrogen and hydrocarbons that hides the surface. The Huygens probe that has successfully landed on the surface as part of the Cassini mission has generated stunning pictures and data to allow more informed speculation about the possibility of primitive life forms since nitrogen and hydrocarbons could provide the materials to make amino acids. There is much more to be discovered about Saturn's other moons although some are thought to keep the rings in order (shepherding) such as the tiny Pan, Atlas, Prometheus and Pandora. As this book is being written, Cassini is returning staggering images of Saturn's moons as well as of the rings and Saturn itself. More about Saturn's moons later.

Uranus was known to have just five moons until it was visited by the Voyager 2 mission. Over 20 have now been identified. The probe passed just 3,000 km above Miranda, giving a resolution of images of an amazing 600 meters. To quote from Sir Patrick Moore's *Data Book of Astronomy* – "the landscape is amazingly varied, and there are several distinct types of terrain: old, cratered plains, brighter areas with cliffs and scarps, and ovoids or coronae, large trapezoidal-shaped regions; one of these, Inverness Corona, was nicknamed the Chevron, while another, Arden Corona, was the Race-track. The cliffs may tower to 20 km; large craters are lacking, but there are fault valleys, parallel ridges and graben [strips of land that have subsided between faults] up to 15 km across". Titania and Ariel are among the larger satellites whose surfaces are covered with a mixture of craters and valleys stretching for hundreds of kilometers.

Neptune has eight satellites, the largest of which is Triton. At 2,700 km in diameter it is the only large moon in the Solar System to circle a planet in the opposite direction to the rotation of the planet. It is the coldest body in the Solar System, reaching a staggering minus 235 degrees Celsius (–235 °C)! Voyager 2 revealed the surface to have extensive cracks and outpourings of nitrogen gas and dust particles. Triton is another moon with a certain end. It is gradually getting nearer to its host planet and will crash or break into bits to form a ring in 10–100 million years time. Nereid, the outermost satellite of Neptune, is the moon with the most eccentric orbit in the Solar System, having an eccentricity of 0.75. At its closest it is 1.35 million kilometers and its furthest 9.62 million kilometers.

Pluto, the outermost planet, has just one satellite but it is half the size of the planet itself. So close are the sizes of the two bodies that the system is often referred to as a double planet. They are gravitationally captured, always having the same face pointing to each other.

Other planets? Who knows how many more planets might be discovered far out beyond Pluto. At the time of writing this book two significant lumps of rock have been found and are referred to as planets. Quaoar has a diameter of 1,250 kilometers and resides within the Kuiper belt of a myriad icy planetesimals. Sedna is somewhat larger at 1,400 km in diameter and occupies a highly elliptical orbit, 76 AU from the Sun at its closest and 880 AU (130 billion kilometers) at its furthest. Whether larger ones with moons will show themselves remains to be seen.

Earth's Moon is discussed at length elsewhere but do we have just the one? Some "quasi-moons" have been discovered recently whose orbits are set round

Table 12.1. Approximate diameter ratios of moon and host planet

Moon	Diameter (km)	Planet	Diameter ratio	Features
Moon	3470	Earth	4	Largest compared to planet except Charon
Phobos	27	Mars	252	Closest orbit
Deimos	15	Mars	453	Smallest moon in Solar System until recent discoveries
Io	3650	Jupiter	38	Most volcanically active
Europa	3130	Jupiter	45	
Callisto	4806	Jupiter	29	Most heavily cratered
Ganymede	5262	Jupiter	27	Biggest in Solar System
Titan	5150	Saturn	23	Second largest moon in Solar System
Tethys	1058	Saturn	113	
Dione	1120	Saturn	107	
Rhea	1528	Saturn	79	
Iapetus	1460	Saturn	82	
Miranda	470	Uranus	106	Most studied of Uranus moons
Oberon	1523	Uranus	33	
Titania	1578	Uranus	32	
Ariel	1158	Uranus	43	
Umbriel	1169	Uranus	43	
Triton	2700	Neptune	19	Only large moon to circle planet in opposite direction of rotation of host planet. Coldest known body in Solar System: −235 °C
Charon	1270	Pluto	2	Largest in relation to planet. Only moon and planet to have same face permanently to each other

the Sun but also pass close to the Earth. However, they are no more than wayward asteroids that cross Earth's path and may be trapped in orbits caused by the gravitational attraction of both Sun and Earth. One has been named Cluithne, is about 5 km wide and takes 770 years to complete an orbit, including an apparent horseshoe-shaped path around the Earth.

It is interesting to view the order of moons with regard to their absolute size and compare them with their host planet's relative size(tables 12.1 to 12.3). While Ganymede is the largest of the moons, it comes down the list when compared to its host Jupiter, although tiny Deimos maintains its position at the bottom of the league on both counts. Some superlatives and notes have been added to indicate the extreme conditions existing among the Solar System satellites.

Jupiter

While it is necessary to decrease the brightness of the image of Jupiter for a satisfactory picture, the opposite applies to capturing images of its four main moons,

Table 12.2. Approximate diameter ratios of moon and host planet arranged in order of moon diameter

Moon	Diameter (km)	Planet	Diameter ratio	Features
Ganymede	5262	Jupiter	27	Biggest in Solar System
Titan	5150	Saturn	23	Second largest moon in Solar System
Callisto	4806	Jupiter	29	Most heavily cratered
Io	3650	Jupiter	38	Most volcanically active
Moon	3470	Earth	4	Largest compared to planet except Charon
Europa	3130	Jupiter	45	
Triton	2700	Neptune	19	Only large moon to circle planet in opposite direction of rotation of host planet. Coldest known body in Solar System: −235 °C
Titania	1578	Uranus	32	
Rhea	1528	Saturn	79	
Oberon	1523	Uranus	33	
Iapetus	1460	Saturn	82	
Charon	1270	Pluto	2	Largest in relation to planet. Only moon and planet to have same face permanently to each other
Umbriel	1169	Uranus	43	
Ariel	1158	Uranus	43	
Dione	1120	Saturn	107	
Tethys	1058	Saturn	113	
Miranda	470	Uranus	106	Most studied of Uranus moons
Phobos	27	Mars	252	Closest orbit
Deimos	15	Mars	453	Smallest moon in Solar System until recent discoveries

Table 12.3. Approximate diameter ratios of moon and host planet arranged in order of diameter ratio

Moon	Diameter (km)	Planet	Diameter ratio	Features
Charon	1270	Pluto	2	Largest in relation to planet. Only moon and planet to have same face permanently to each other
Moon	3470	Earth	4	Largest compared to planet except Charon
Triton	2700	Neptune	19	Only large moon to circle planet in opposite direction of rotation of host planet. Coldest known body in Solar System: −235 °C
Titan	5150	Saturn	23	Second largest moon in Solar System
Ganymede	5262	Jupiter	27	Biggest in Solar System
Callisto	4806	Jupiter	29	Most heavily cratered
Titania	1578	Uranus	32	
Oberon	1523	Uranus	33	
Io	3650	Jupiter	38	Most volcanically active
Umbriel	1169	Uranus	43	
Ariel	1158	Uranus	43	
Europa	3130	Jupiter	45	
Rhea	1528	Saturn	79	
Iapetus	1460	Saturn	82	
Miranda	470	Uranus	106	Most studied of Uranus moons
Dione	1120	Saturn	107	
Tethys	1058	Saturn	113	
Phobos	27	Mars	252	Closest orbit
Deimos	15	Mars	453	Smallest moon in Solar System until recent discoveries

Callisto, Io, Europa and Ganymede. They are mere pinpricks of light compared to the planet itself. However, the picture, of necessity, must contain both planet and moons and we have to accept a grossly overexposed Jupiter.

Select an eyepiece of around 25 mm and focus on Jupiter. Its moons will be clearly seen (figure 12.1). Position the camera. Spot-metering is not required. Select the brightness control and set it to +2.0. Take a picture using, as always, shutter delay. Following the delay the camera will take a second or so to snap the picture. Note the number and position of the moons (satellites) as the image taken briefly shows itself after capture. This overall view is necessary because it is easy to miss one at a distance from the planet when zooming in to obtain the best picture. Now zoom in to allow all moons to just fill the frame.

At the time of this picture all moons were nicely spread (figure 12.2). The picture below also shows a good spread but in a completely different orientation (figure 12.3).

In the gallery of six pictures shown, figures 12.4a-f, the top two were taken just one hour apart, indicating just how quickly some objects in the sky can change. One moon has appeared from behind Jupiter in the second picture. It is often

Figure 12.1.

Figure 12.2.

Figure 12.3.

rewarding to take the same photograph at the beginning and end of a session. There is an almost infinite number of patterns of the moons available to the astrophotographer.

Saturn

With the modest 3 megapixel digital camera a little more persistence is required to enjoy the luck of observing at a time of extremely good seeing (figures 12.5 and 12.6).

The two pictures clearly show two major spots and a couple more barely visible – the second one being marginally affected by a light breeze blowing at the time. However, figure 12.7 reveals more satellites at a time of great seeing. In all of these images care must be taken to differentiate between Saturn's moons and background stars – not always too easy but assisted by reference to computer night-sky simulation programs. This image was excitingly recorded within a few hours of the actual landing of the Huygens probe on the moon Titan. Couldn't quite make out the actual probe on this picture though!

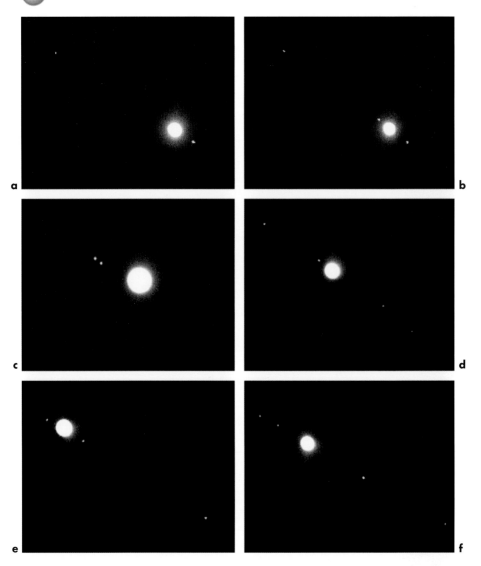

Figure 12.4.

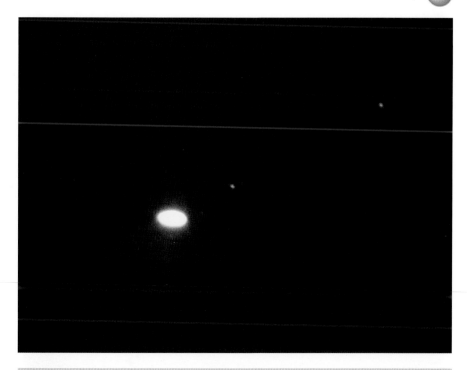

Figure 12.5.

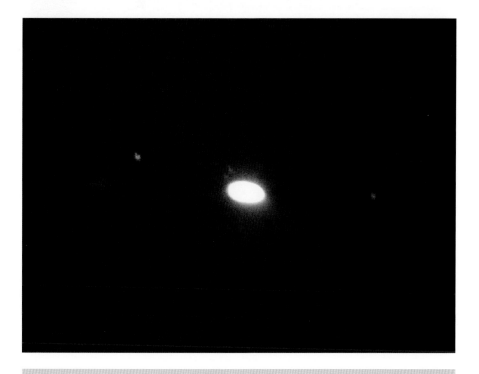

Figure 12.6.

Figure 12.7.

CHAPTER THIRTEEN

The Planets

Overview

Of course we can't record images like this one (figure 13.1) of Saturn with our family camera. It was taken by NASA from the Cassini–Huygens spacecraft when

Figure 13.1.

Figure 13.2.

just around 70 million kilometers away on 9 February 2004. The details of the ring systems are remarkable.

Neither can we easily take pictures as in figure 13.2. This was created when the author combined individual frames from a movie taken with a converted webcam attached to the telescope described in this book – although with no further image processing. But we can still obtain enough detail to be able to recognize some features such as Saturn's rings. Again, very satisfying having accumulated pictures using the inexpensive equipment described.

Eventually, progression to dedication to the hobby of astrophotography will drive you to seek ideal seeing conditions and waiting months, if not years, to capture that one image you have been waiting for. However, following the theme stated earlier of making do with the opportunities available to most of us the following pictures were obtained in a light-polluted area, from a small patch of sky between houses and hills and over a couple of months as winter turned into spring.

Equipment

The large size of the Moon makes it a very forgiving feature as far as taking pictures is concerned but photography of small objects the apparent size of planets is much more demanding. The optical performance of the Schmidt–Cassegrain telescope is directly related to the accuracy of its collimation, the alignment of its optical system, and it is therefore worth spending a little time to check this. The technical literature dictates that you focus on a bright star and make sure your telescope drive tracks it to maintain the view. Choose a medium to high power eyepiece and bring the star in and out of focus. If there is a systematic skewing of the star then recollimation is needed. Adjust the screws on the front of the telescope to move the secondary mirror. Consult the telescope manual or other technical book or a colleague for full guidance. Testing and adjustment of the collimation can be made much easier. Look at these two pictures (figures 13.3A

Figure 13.3.

and 13.3B). The first shows a garden with a fence. The second reveals a shiny round-headed screw driven into the fence. Viewing the screw from around 15–20 meters away with the telescope simulates a good approximation to viewing an image the size of a bright star in the sky. The advantages are clear. The observations and adjustments can be carried out in comfort, with the scope horizontal and plenty of light to see exactly what you are doing.

Jupiter

Compared to the Earth, Jupiter is massive indeed; its mass is greater than all the other planets put together. However, it is quite unlike the Earth. All planets circulating the Sun in orbits with radii less than Jupiter's are small and rocky but although it has a small rocky core (probably) it is almost entirely made up of liquid and gaseous hydrogen and helium. The upper atmosphere is, of course, all we can see and therefore photograph through our telescope. Jupiter is still cooling off from its formation and the heat triggers phenomena ranging from the planet's enormous magnetic field to convection currents in the clouds. Bands of color are a familiar characteristic of its surface and a consequence of high cool updrafts and warm downdrafts. Because of the rapid rotation of the planet, about 10 hours, the bands of clouds circulate at slightly different speeds. Storms are generated within the clouds, the most famous of which is the Great Red Spot that has endured for several hundred years.

Jupiter has yet another generous vision. Of its many moons four are very easily seen and photographed through the telescope and even quite clearly seen through a good pair of binoculars. They are Ganymede, Callisto, Io and Europa. They orbit the planet with a rotation of approximately 1–16 days. Consequently their positions each night are well worth recording as already described.

Certainly, compared to the Moon, Jupiter is a very small spot of light indeed and through the digital camera it looks tiny. Set up the telescope, contraption and camera as described previously. Use a 25 mm lens with a ×3 Barlow or the nearest equivalent combination in your possession. Experiment with other eyepieces also. It may be necessary to use a low-power lens initially to find Jupiter. Focus on the planet this time by moving the camera away from the eyepiece to get your eye in the correct position. Then do not move the eyepiece (as for the Moon) but use the contraption adjustments to allow the camera to hover above the lens. Every opportunity must be taken to squeeze the most out of the equipment to achieve the best focus. Whatever the target a poor focus makes a poor picture.

Although Jupiter appears extremely bright the overall amount of light received by the camera is small, i.e. a spot of light against the black remainder of the screen. The automatic exposure compensation will therefore render the image so bright as to be completely washed out as shown in figure 13.4. Fortunately something can be done about this. First, select the spot-metering feature then put on full zoom. The image of Jupiter just nicely fills the small central square frame you see on the back screen that the camera uses for exposure calculation. If not, proceed as for Saturn – see later. The next step requires a very steady hand. Select

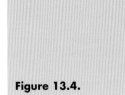

Figure 13.4.

Figure 13.5.

the shutter time delay, then very, very carefully press the shutter release without allowing the image to drift out of the square frame or the wrong amount of exposure will be obtained – washed out again. Remember that the exposure calculation will be made at the time of pressing the button not at the time the picture is taken. By noting the photographic result that appears for a second following capture you will decide whether success has been achieved in maintaining a steady hand. If not – try, try, try again. That is the beauty of this technique – being able to continue taking many pictures and making adjustments accordingly. The picture (figure 13.5)! now reveals some detail of the stripes or bands of the planet.

Processing a Movie

It was mentioned at the beginning that good pictures of the planets can be obtained by taking webcam movies and aligning and averaging each of the individual still frames. The digital camera has a movie facility so the question has to be posed – can the same effect be achieved with it? To some extent yes, although much more challenging! First, consider the resolution or the fine detail it is possible to reproduce. The webcam photo detector typically has an array of 640×480 pixels or picture elements. The higher the number of pixels the better the resolution and the larger the final photograph can be printed. The number of frames captured per second, the duration of capture, the brightness and contrast can be fully controlled. Our digital camera has a more limited versatility. It takes movies in bursts of around 30 seconds at a rate of 15 frames per second at a maximum resolution of 320×240 pixels, i.e. much less than the webcam. Only an outline of the method is described here. To carry out the procedure either seek assistance or, if you are adventurous enough, use a trial and error technique. After all, nothing is lost by trying and the cost of trying with a digital camera is close to nothing.

Set up the telescope alignment, exposure, shutter delay, focusing, etc. already described for Jupiter as above. Select "movie" on the camera menu and press the shutter button. The movie will be accumulated. At the completion of your photographic session download pictures onto your computer in the usual way. Movies take a little longer than stills. Manipulation of the movie is now required to break it down into its component frames that are then fed into a computer program to align them then stack or average all of the accumulated pictures. RegiStax and AstroStack are well-known programs to achieve this. Even better – they are available as free downloads with instructions from the Internet. Unfortunately, an additional step is required for your camera movie. The movie is created in a format possibly incompatible with the processing software available. Fortunately, rescue again comes from the Internet where yet more programs are available to convert your camera MOV (or other) format to RegiStax and AstroStack compatible AVI format. Again, seek advice to carry out this sequence of operations. Figure 13.6 shows one result of all these efforts – a reasonable product although somewhat "pixelated" (blotchy) due to the relatively low resolution of the camera detector. A little simple image filter adjustment (gamma) contained within the camera software supplied has been applied to this image. For your particular camera some of the details such as the type or format of software supplied might be different from the above but the principles of accumulating and processing images will be similar. Therefore, to summarize:

1 Take a movie sequence and transfer to your computer.

2 Download a program to convert your MOV file to AVI, e.g. Artech allows a free trial to assess prior to purchase (artech365.com).

3 Download and open RegiStax (aberrator.astronomy.net/registax) or AstroStack (astrostack.com) or an equivalent program. Input your converted file and align and stack as instructed.

4 RegiStax and AstroStack also allow you to attempt further processing of the stacked images.

Figure 13.6.

Combining Single Images

Another question poses itself. Instead of accumulating many low-resolution pictures (450 from 15 frames per second and 30 seconds exposure) for alignment and stacking why not take several high-resolution "stills" and feed them into a stacking program? After all, the Moon and other Jupiter pictures were taken at a resolution of 1984×1488 pixels that dwarfs those of the movie of 320×240 pixels. Here is an example of such an experiment. Approximately 30 high-resolution pictures were taken of Jupiter. Due to the variation in brightness of the image as the camera's automatic exposure metering facility attempted best value, the images were grouped into bright, dim and really dim to assess best conditions (figures 13.7, a-f).

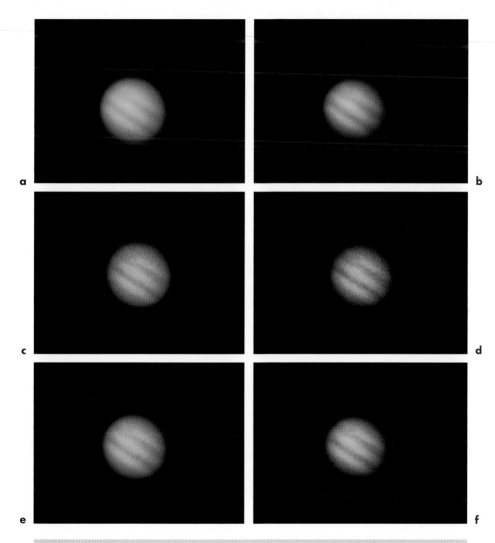

Figure 13.7.

Figure 13.8.

Each group was processed separately and shown on the left with simple gamma and contrast adjustments on the right. Remember that because of the camera orientation all pictures appear upside down in the first instance. Processing all pictures together, without separating for brightness, creates a very washed-out appearance. In addition to the clear bands of clouds, the Great Red Storm spot can now just be seen.

Another feature sometimes available is the passing of moons across the face of Jupiter – known as a transit. In this single-frame shot, figure 13.8, is just discernable the beginning of a transit of a moon – left edge of the image (must do better!). The more pictures taken the more likely opportunities arise to capture a variety of events and features.

Saturn

Surely Saturn must be the most beautiful object available to the amateur astronomer and especially to the beginner. Everyone remembers their first view of this awe inspiring planet. The majestic rings are a feature of beauty and, although it is now known that other planets have rings, Saturn's are the only ones visible to the amateur backyard astronomer. Relative to the Earth the angle of the plane of the rings changes on a 15 year cycle and because the rings are so thin, possibly a very few kilometres, edge-on they cannot be seen. The rings are essentially composed of three bands (although high-resolution equipment can record many more) separated by a gap between the outer A and middle B ring, named the Cassini division. At times of excellent atmospheric viewing conditions the gap can just be photographed. Like Jupiter the visible surface of the body of Saturn has bands but much less prominent although it is still possible to capture their image.

Saturn poses a greater challenge than Jupiter to the astrophotographer. The image is much smaller but is a very bright spot. Focus on the planet and set shutter delay (and of course flash off), reduced brightness (−2.0), spot-metering

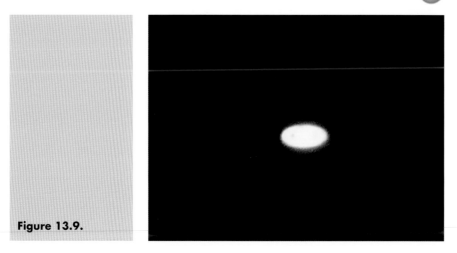

Figure 13.9.

and full zoom. Take a picture and what do you get? An overexposed white blob (figure 13.9). The cause is the size of the image compared to the size of the inner frame for spot-metering. It doesn't fill it. Therefore the camera detects much background darkness within the frame and increases the exposure. The author has discovered a trick – a cheap solution to the problem.

Remember that the camera sets the exposure rating at the time of pressing the shutter button. Then approximately 10 seconds later (the camera's fixed shutter delay time) takes the picture. It is necessary therefore to trick the camera into thinking a bright object is the target. Shine a cheap and not too bright torch directly into the telescope with one hand and press the shutter release with the other (figures 13.10 and 13.11). Remove the torch and the exposure has been set for a brighter image. It takes a little practice and some images are too bright, some too dark but some are just right (figure 13.12). Here we can clearly see the rings and just beginning to see a hint of two different bands within them. Quite an achievement for our family camera and not-wildly-expensive telescope. But

Figure 13.10.

Figure 13.11.

Figure 13.12.

can improvements be made? Two additional techniques were used for Jupiter – stacking individual pictures and processing a movie.

Processing a Movie

The torch is not necessary for this process since the 15 frames per second naturally reduces the exposure. Once the movie has been captured repeat the opera-

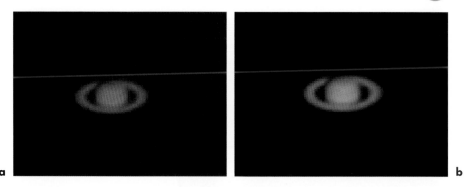

a   b

Figure 13.13.

tion of loading onto the computer, converting the file format from MOV (or equivalent for your camera) and inputting to a stacking program such as AstroStack or RegiStax. Two examples are shown here in figures 13.13a and 13.13b. Some gamma, brightness and contrast adjustments have been applied but with assistance even better images can be produced.

Combining Single Images

The picture shown in figure 13.14 is the result of stacking several single-shot images with a little gamma, brightness and contrast adjustment. Zones in the

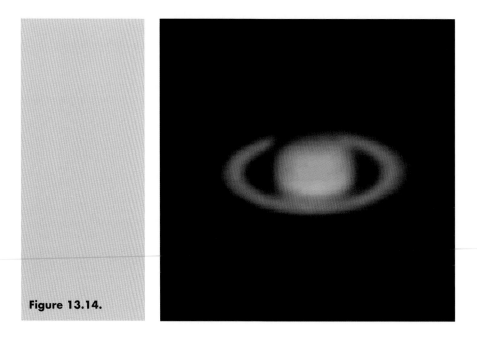

Figure 13.14.

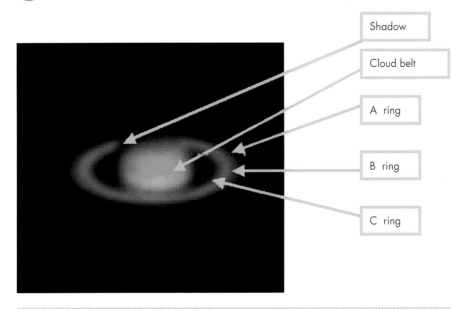

Shadow

Cloud belt

A ring

B ring

C ring

Figure 13.15.

rings and some structure on the body of Saturn itself are just visible. Although still not as good as webcam pictures these images represent very satisfying results for the amateur with a digital camera. A key to what can be seen is given in figure 13.15. Although the ring structure is exceedingly complex, comprising a multitude of thin rings, the three main bands A, B and C can be distinguished and a cloud belt on the planet surface. A shadow is cast across the rings at the far side.

In addition to obtaining good photographs don't forget to enjoy the wondrous view through the telescope itself.

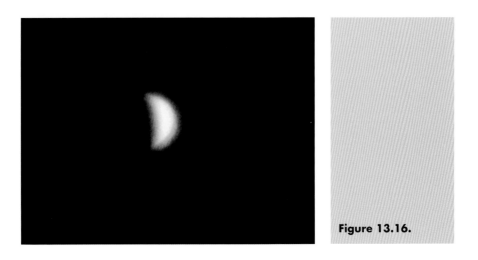

Figure 13.16.

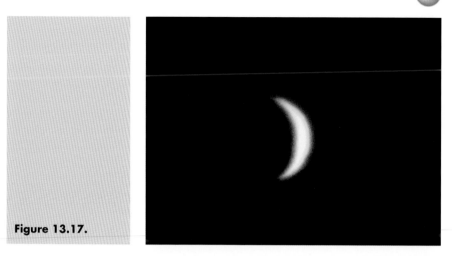

Figure 13.17.

Venus

Apart from the Sun and Moon, Venus, at its best, is the brightest object in the sky and in terms of size and mass is similar to the Earth. However, in all other respects it is very different. From a photographic point of view the surface is unavailable due to the dense clouds covering the planet. Although the surface of Venus appears quite featureless it does show phases like our Moon, first observed by Galileo in 1610, and is therefore well worth capturing through the telescope (figures 13.16 and 13.17).

At the time these pictures were taken – with 25 mm and ×2 Barlow lenses – the planet almost, but not quite, filled the selection square for automatic exposure compensation on full zoom. It was therefore necessary to select –2.0 on the brightness setting to achieve an acceptable picture. Since further efforts could not

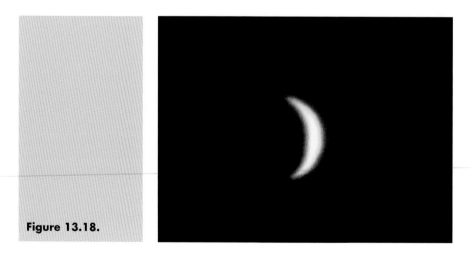

Figure 13.18.

reveal additional detail other techniques to reduce the brightness were not required (figures 13.18 and 13.19).

Figure 13.19.

CHAPTER FOURTEEN

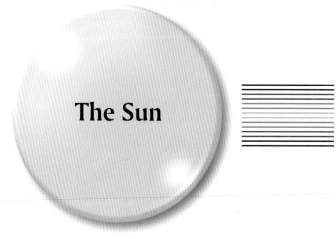

The Sun

Overview

It is taken for granted that astronomical viewing means standing in the cold and dark at night, often in a muddy field or other less than comfortable environment. How nice also to have, as part of the hobby, times when only sunny days will do. The Sun is a fascinating subject and although there are limitations as to what can be seen using amateur equipment there are still plenty of pictures to capture.

"Where is our nearest star?" So often this question can lead newcomers to astronomy, especially children in school, to guess at the tiny pinpricks of light in the night sky instead of pointing to our Sun (figure 14.1). It is a main-sequence star so-called because it is doing what most stars do during the main part of their life span. It converts hydrogen to helium to produce its light and heat. It is one of 100,000 million stars making up our galaxy of the Milky Way. It is a good job it is about 150 million kilometers away – the heat it generates gives rise to temperatures of 5,000 degrees Celsius at the surface (photosphere) rising to 1 – 5 million degrees further out into the upper atmosphere (chromosphere and corona). The very dense core is estimated to have a temperature of 15 million degrees. Heat is transported from the center by radiation and through the outer layers by convection. And what causes all this heat? The Sun is fuelled by the thermonuclear reaction of hydrogen atoms fusing together to produce helium plus energy: 600 million tons of hydrogen are converted into helium *every second*! However, the energy produced at the center takes a very long time, hundreds of thousands of years, to reach the surface. Sunspots frequently appear as dark patches since they are slightly cooler than the surrounding areas and can occupy areas as large as thousands of kilometers across. Sunspots are a result of the tortuous and

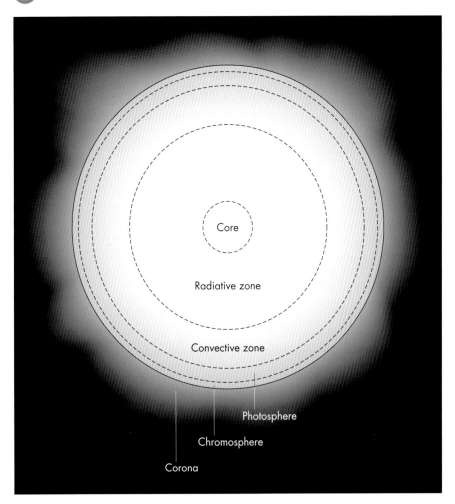

Figure 14.1.

twisted magnetic field and since the Sun rotates approximately every 27 days it is interesting to see how the spots have changed once they disappear then come into view again after about two weeks. Using the correct telescope filters, violent eruptions known as prominences can be seen to squirt out to hundreds of thousands of kilometers.

At 1,400,000 km in diameter, the Sun is huge compared to the Earth and even comprises more than 99.99% of the mass of the Solar System. Gravity of the enormous mass of hydrogen and helium – in addition to smaller quantities of other elements such as oxygen, carbon, nitrogen, silicon and iron – holds the ball of gas together while the heat attempts to blow it apart. The battle will become unequal in about 5 billion years time, the Sun expanding as it gets brighter and hotter and growing into a red giant, finally contracting to a white dwarf. It is about halfway through its life.

Sunspots

It cannot be repeated too often to take extreme care when viewing the Sun. Before looking through the telescope check that the correct solar filter is securely in place. Cover the viewfinder. Don't leave the scope unattended. Check every time. Also see the safety warnings in early chapters.

With an inexpensive solar filter many features can be seen but *not* the prominences and flares streaming from the surface. Those pictures will be available to you later when investing in an expensive hydrogen-alpha filter (although prices are reducing rapidly). Purchase a filter, which may be coated glass or polymer film, from a knowledgeable supplier. Sunspots resulting from magnetic disturbances appear as dark areas that may be very small or immensely large. On occasions there may appear to be none at all. By keeping a constant watch for just a minute each day (even from indoors through your window!) you will be alerted to significant viewing opportunities worthy of transferring the telescope outside. If you wait until you are alerted through the media or magazines it could be too late.

There came such an opportunity during October 2003 when a magnificent display of spots appeared. First here is a picture of the "blank" Sun (figure 14.2). Certainly no spots in sight here.

Essentially, on the 26 October three main Spot groups appeared – one small and two large (figures 14.3 and 14.4).

The following day another appeared. This one was also very large and an intriguing shape (figure 14.5).

And the next day the shapes had changed significantly (figure 14.6).

And yet more changes were seen on the 31 October (figure 14.7).

And a couple of days later …(figure 14.8)

Figure 14.2.

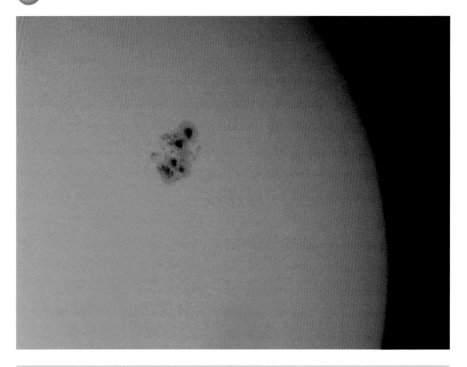

Figure 14.3.

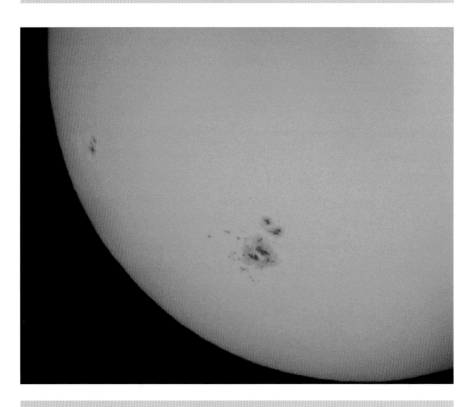

Figure 14.4.

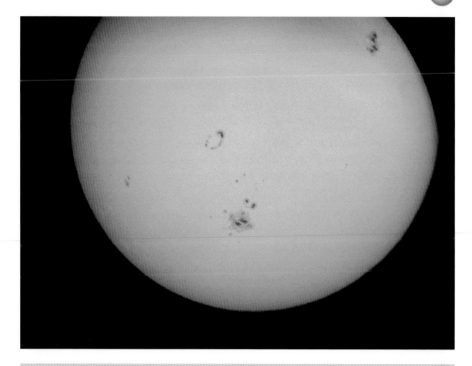

Figure 14.5.

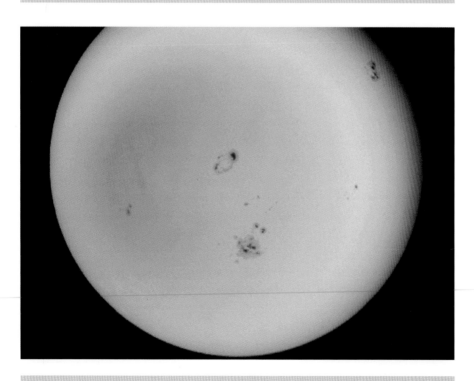

Figure 14.6.

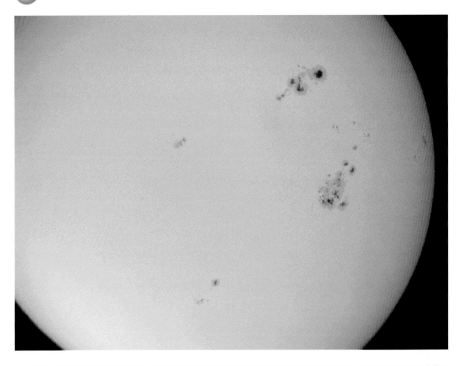

Figure 14.7.

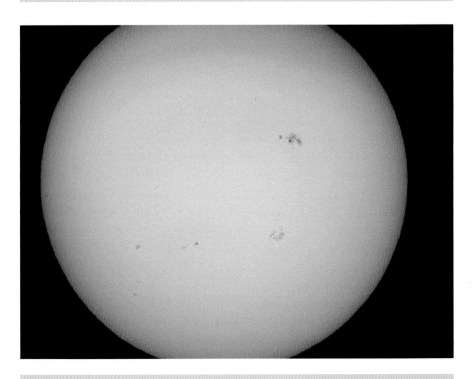

Figure 14.8.

Having accumulated 60–100 pictures at each session, several were found to be worthy of further display (figure 14.9, a-f).

Notice that some images are sharper than others. As for the eclipse, often less than top quality must be accepted to maintain a record of continuity of the event.

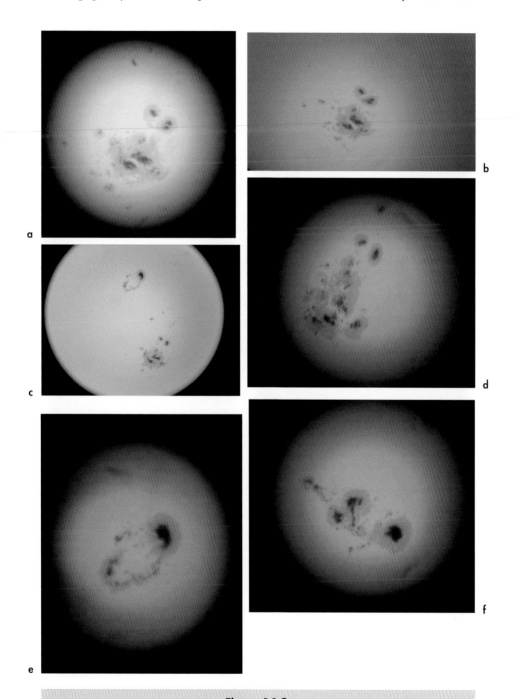

Figure 14.9.

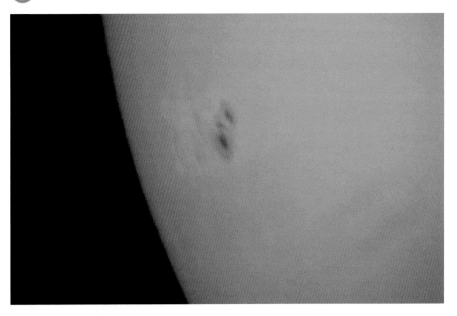

Figure 14.10.

Seeing conditions are out of your control. Notice also the white surround in figure 14.10 that is a slightly hotter region of incandescent gases. These "faculae" most often occur near sunspots and are especially visible near the limb (edge) of the Sun's disk. Sunspots themselves, of course, are extremely hot and bright but less so than the surrounding furnace, so appear relatively dark.

As the Sun rotates approximately once every 27days (relative to the Earth) it provides plenty of opportunity (weather permitting) to monitor spot development across the face. It is very exciting waiting to observe differences in the patterns of large spots as they reappear a couple of weeks later.

Eclipses of the Sun

A total solar eclipse is one of Nature's greatest sights. It happens whenever the New Moon passes directly between the Sun and the Earth. An eclipse is not seen at every New Moon because the Moon's orbit is inclined at about 5 degrees to the ecliptic. In most months the Moon passes above or below the Sun instead of across it. Solar eclipses can be total, partial or annular depending on how much of the Sun is covered. A total eclipse may last a mere second or longer but only up to about seven and a half minutes. It is pure coincidence that the apparent size of the Sun and Moon are about the same. The Moon is roughly 400 times smaller than the Sun but happens to be about 400 times closer. In fact, since the Moon formed, its orbit has been slowly moving away from us and will appear to be steadily shrinking. Eventually it will orbit too far from Earth to ever completely

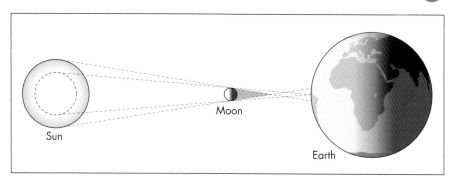

Sun

Moon

Earth

Figure 14.11.

cover the Sun, thus removing the possibility of total solar eclipses. Due to the rotation of the Earth and movement of the Moon, the lunar shadow, or umbra, traces out a curving path on Earth. The umbra barely reaches Earth so the path is relatively narrow. Where the Moon is directly overhead the path will be at most 270 km wide. Total eclipses will therefore only occur rarely at any particular place, although in most years there are one or two eclipses somewhere on Earth.

Consider the reason for an annular eclipse (figure 14.11).

The diagram shows that the Sun's rays from the edges form a cone of shadow that does not reach the Earth. Therefore only the amount of the Sun shown by a dotted circle is blotted out by the Moon, forming a "ring of fire".

The weather notoriously interferes with eclipses but this image captured was just sufficient to prove that it happened for me – the annular eclipse of the Sun on 31 May 2003 (figure 14.12). The shape of the Moon can be seen, leaving the Sun at bottom left. Weather precluded vision of the Sun for many in Scotland sitting in the path of the full eclipse but for some elsewhere in Britain clouds parted to see the partially eclipsed sunrise.

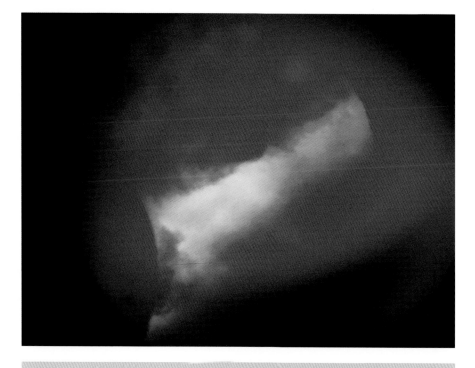

Figure 14.12.

CHAPTER FIFTEEN

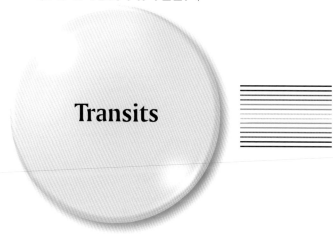

Transits

Overview

A transit is the phenomenon of one relatively small astronomical body passing across the face of another larger one. The four Galilean moons regularly cross the face of Jupiter, but the term is most frequently used to refer to the passage in front of the Sun of the inferior planets Mercury and Venus. Although to the uninitiated staring at a little black dot passing a big yellow one comes close to the excitement of watching paint dry, to the keen astronomer and those with enquiring minds and the ability to be amazed, a transit is a truly stunning event. Not least among the reasons is the rarity. On average there are 13 transits of Mercury each century but more than a century separating each pair of those of Venus. One might expect transits by the two planets to occur at every inferior conjunction when Venus and Mercury pass between Earth and the Sun. However, their orbits do not lie exactly in the plane of the ecliptic and so they usually pass just above or below the Sun.

The main points occurring during a transit are the contacts illustrated in figure 15.1. The exciting first contact is when the planet first touches the Sun. A cheer will often be heard from all observers in the party. History is being made. Many will note the exact time, to be used later for calculations of the Sun's distance from Earth. Second contact arrives at the point when the planet is just fully on the face of the Sun. Is the "black drop" seen – the apparent elongation of the blackness as the planet outline just fully enters that of the Sun? Is there a ring around the planet due to an atmosphere? Once again times are noted. All are glued to their telescopes or have their heads in cardboard boxes to provide the shade necessary for their computer monitors or indirect viewing screens. For the next

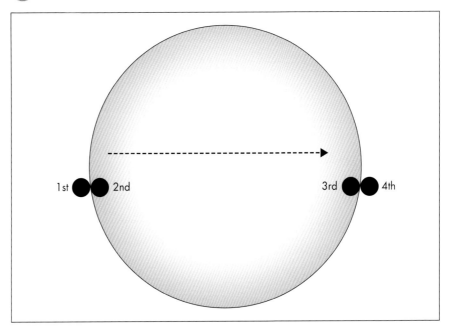

1st   2nd   3rd   4th

Figure 15.1.

several hours the very black dot slowly creeps across the bright disk of the Sun. Dare anyone to miss these once in a lifetime pictures. Contact three occurs when the dot reaches the limb of the Sun for egress. Again eyes are strained searching for black drops and rings. A sadness descends on the party as the planet begins its departure to say farewell at contact four. Everyone calls out in an attempt to define the precise moment when the transit is over and the final timing is noted.

Mercury

The transit of Mercury of 7 May 2003 was visible from Europe, Africa and Asia. The weather plays a crucial role for observations, of course, but from the UK many places had clear skies although first contact was clouded out for some.

Mercury is a mere 1/158 of the Sun's apparent diameter and therefore appears as a very small dot indeed. However, its sheer blackness stood out remarkably compared to the sunspots that also presented themselves on the day. While viewing it is interesting to recall exactly what is being seen. Part of the thrill is that Mercury's orbit is so close to the Sun that it never strays more than 28 degrees from the blinding light of the Sun and therefore is less often available to camera and telescope. And it is small at a diameter of 4,878 km. Most of our knowledge has been accumulated from the Mariner 10 probe that arrived at the planet in 1974. Mercury has a very large iron core, over 70% of the total weight of the planet, the remainder being rocky mantle and crust. Circling close to the Sun,

Figure 15.2.

having a very elliptical orbit and slow rotation leads to a wicked temperature range. Where the Sun is overhead at perihelion the temperature rises to +127 degrees Celsius and down to –183 degrees at night. It has virtually no atmosphere and a small magnetic field – about 1% of that of the Earth. The surface is similar to that of the Moon, highly cratered with scarps but without the Moon's dark lava flows. Only half of the planet was photographed but features seen such as craters, plains, mountains, valleys, scarps, ray systems and ridges are very likely to be repeated on the unmapped side.

The equipment already described for collecting solar pictures is superbly suitable for capturing a great many images in sequence for the infrequent transits of Mercury or Venus across the Sun. The day of 7 May 2003 was lovely and sunny and ideal for observation of a Mercury transit. In figure 15.2 Mercury is the small black dot to the left of the picture shortly after beginning its transit. Thin cloud precluded observation of first and second contacts. The central and right dark smudges are sunspots. Notice how black Mercury is compared to the sunspot in figure 15.3 (left). Figure 15.3 (right) shows Mercury well on its way.

And as it exits the disk completely (figure 15.4a-f).

There is plenty of time during a transit to capture other features of interest. In figure 15.5 the faculae, already described, can clearly be seen as isolated patches as well as surrounding the darker sunspots.

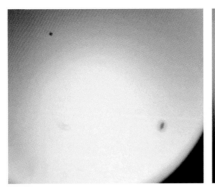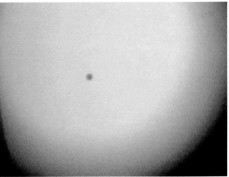

Figure 15.3.

Venus

What terrific anticipation there was for this unique occasion – unique from the point of view that nobody alive has seen one and will never see another, unless travelling some distance to view the second of the pair. The last occurred in 1882, 122 years ago. Venus transits occur in pairs the next being in June 2012 although this event will be centered on the Pacific Ocean and will not be visible in Europe. The first ever to see a Venus transit were two amateur English astronomers Jeremiah Horrocks and William Crabtree in 1639. Sufficient became known to prepare for the next occurrence 130 years later when James Cook (later Captain Cook) was sent to Tahiti to hopefully experience clear skies and record accurate timings.

What sort of a planet is Venus? It is often referred to as Earth's twin. It is indeed close to us in terms of diameter – Earth 12,756 km, Venus 12,104 km – and mass, but what a difference in other respects. It has little or no magnetic field (figure 15.6). For man it is a most hostile place as determined by a few dedicated probes and some taking notes as they passed by on their way to other parts of the Solar System. The surface temperature is a searing 467 degrees Celsius night and day, easily hot enough to melt lead. The atmosphere is predominantly carbon dioxide, hence the runaway greenhouse effect, and the pressure around 90 times that of the Earth. Corrosive sulfuric acid pervades the clouds. Venus rotates only once in 243 Earth days, which is longer than its year of 225 days, and spins in the opposite direction to most other bodies in the Solar System. This hints at the possibility that it was struck by a lump of space debris early in its history. It suffers from very high-velocity winds around 350 km/hour at cloud-top level but far less at the surface. Volcanoes and lava flows are everywhere.

June 8 2004. Everyone was crossing their fingers hoping for clear skies. The finger-crossing worked – it was a beautiful day in Selsey. A transit party had been arranged by Sir Patrick Moore in his garden and I was proud to have been invited there with him (figure 15.7). There were about a dozen telescopes plus gadgets of all sorts ranging from very expensive scopes with hydrogen-alpha solar filters to

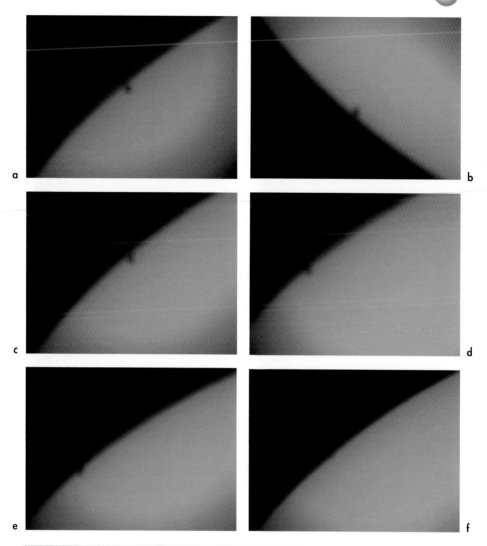

Figure 15.4.

inexpensive ones projecting the Sun's image onto screens inside boxes to provide shade for viewing. Two youngsters (figure 15.8) view correctly, safely and enthusiastically. No one needed reminding of the absolute requirement of safety – correct filters or indirect viewing only.

Right on cue the black dot slowly edged its way onto the solar disk around 6:20 am (5:20 GMT) about 12 degrees above the north-eastern horizon. The challenge for all was to announce when they had decided the exact moment of entry or ingress. It was important that the moment of decision was accompanied by logging the precise time of this first contact to the nearest second (to be explained later). It is worthy of note that the Sun was almost completely devoid of sunspots

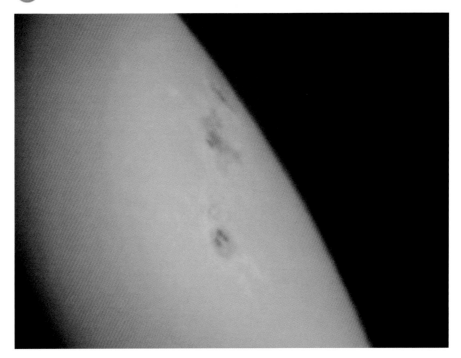

Figure 15.5.

and although that removed any tiny possibility of anyone confusing black dots it would have made a somewhat more interesting picture. For the first time, modern astronomical equipment was available for monitoring the transit presenting the possibility of looking for two features discussed. First, a light ring appearing round the planet as it crosses into the Sun's disk after first contact due

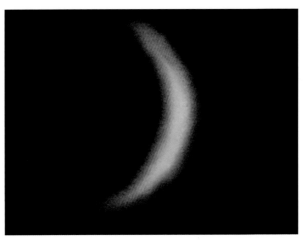

Figure 15.6.

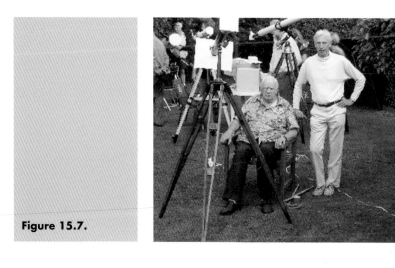

Figure 15.7.

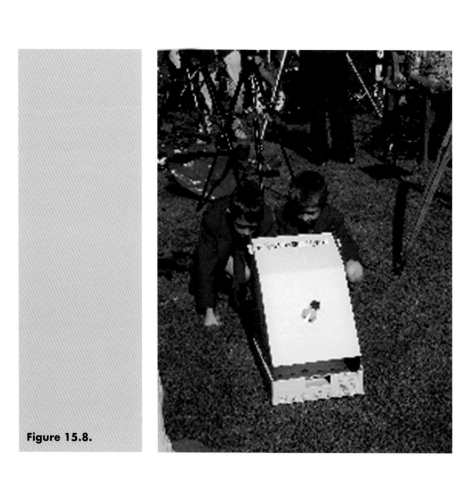

Figure 15.8.

Figure 15.9.

to the refraction of light from the Sun through Venus' atmosphere. Secondly, as the planet just enters the Sun's disk completely, a "black drop" appears to join the limbs of planet and Sun. Is it due to the planet's atmosphere or just an optical effect? I am sure conclusions will be drawn following analysis of the many images recorded.

Photographically and telescopically there is a great flurry of activity between first and second and third and fourth contacts but there are many hours in between. That provides a great opportunity to inspect and discuss the equipment and techniques of other astronomers. Poking a digital camera down other telescope eyepieces is another great opportunity and shown here is a photograph through a hydrogen-alpha filter, although some allowance for the quality must be made since the camera could only be hand-held and hand-steadied (figure 15.9). Using the computer to overenlarge the image some Sun prominences can just be discerned. Prices of these expensive filters are slowly coming down and may well be within reach of many more amateurs soon. Another shot shown here (figure 15.10) is that taken, hand-held, through the smaller ETX70 with a glass solar

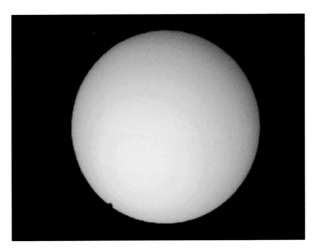

Figure 15.10.

filter. It is also a time to check the screws on telescope pivots and joints. Although the telescope sits almost level at first contact, by the time of third and fourth contacts it will be pointing almost vertically and the contraption and any other attachments will slip if they get the chance!

As Venus approaches egress, the last two contacts, excitement is gradually replaced by a nostalgic sadness knowing that we will never see the like again. A sorrowful "goodbye Venus" can be heard from the party as our black-dot companion of the last six hours fades and purposefully continues its travels round the Sun.

Here are some of the images captured. Firstly the moment of decision. Is it there yet? Has first contact arrived? (figures 15.11 and 15.12)

A hugely blown-up portion of figure 15.12 confirms that Venus is on its way (figure 15.13).

About halfway onto the Sun's disk (figure 15.14) – and – is this second contact? (figure 15.15) Now enjoy the majestic progress of Venus and the making

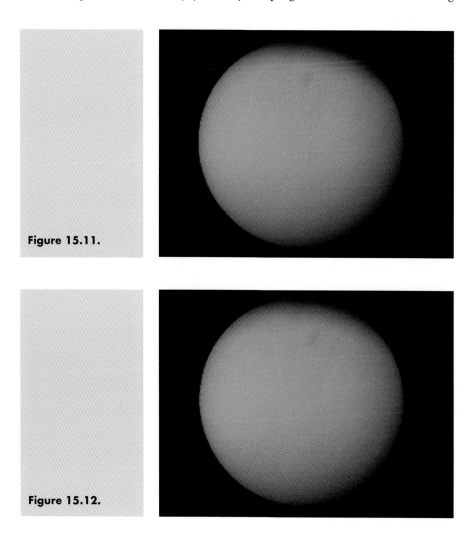

Figure 15.11.

Figure 15.12.

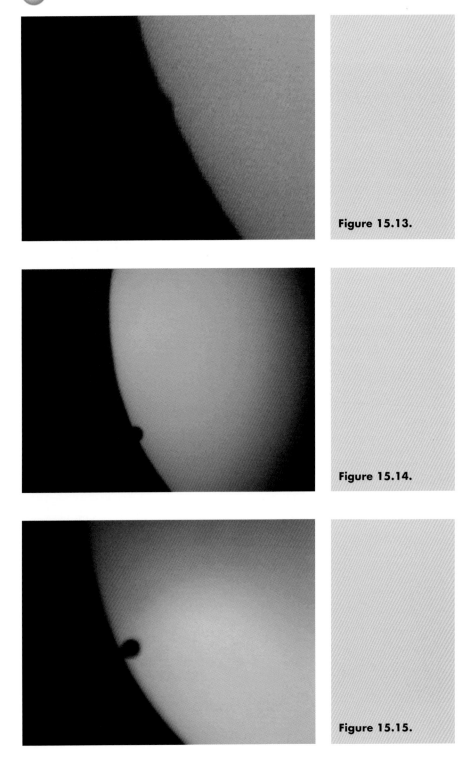

Figure 15.13.

Figure 15.14.

Figure 15.15.

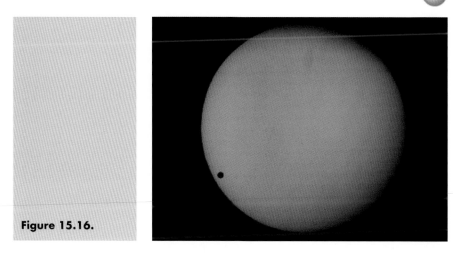

Figure 15.16.

of modern history. Note how, as for Mercury, the spot is very black. A tiny sunspot can just be discerned at the Sun's center (figure 15.16).

Plenty of transit remaining so time to play games! Extra magnification! And more. (figure 15.17) But there is no point of course. Not even the greatest of telescopes can penetrate the thick Venusian atmosphere let alone inexpensive astronomical equipment and a family camera!

Note how the coloration has changed from the orange hue of early pictures taken at a low angle to the gray of a more vertical shot through less atmosphere and lower refraction. (figure 15.18 left)

Just about halfway now (figure 15.18 right) … and … nearly there (figure 15.19 left)! Is this third contact (figure 15.19 right)? Perhaps a major enlargement will help (figure 15.20), since an error of a second or two will cause a big difference in the estimated distance of the Sun (see below). I have to confess that, try as I might, I could not detect a black drop nor a ring surrounding the planet.

Just about half the planet's diameter away from fourth and final contact (figure 15.21 left). Then the last possible trace of our friendly black smudge (figure 15.21 right) to declare final contact and note the exact time. Goodbye dear Venus.

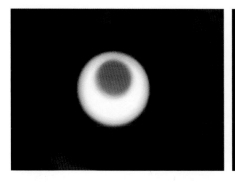

Figure 15.17.

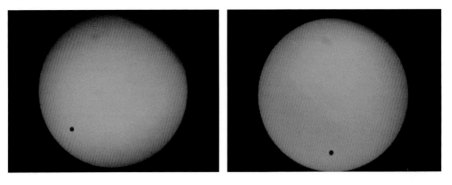

Figure 15.18.

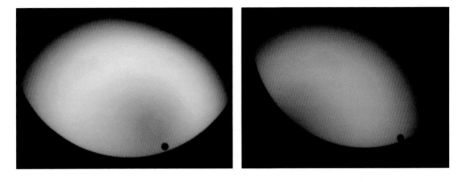

Figure 15.19.

What can be achieved after the event? The priority is certainly to plough through the hundreds of photographs searching for the significant moments of all four contacts. Edmund Halley, of Halley's comet fame, in 1716 was the first to suggest that the distance of the Earth from the Sun could be calculated from observations of a transit of Venus across the face of the Sun. It would require measurement of the geometrical parameter of parallax, the angular difference between an object's direction as seen from two points of observation. In essence two observers should be as far apart on Earth as possible and record exact times of positions of Venus as seen from their vantage point. Halley knew that he would not live to see the event but others prepared and travelled to distant parts of the world. Hence the travels and tribulations of James Cook and others. With modern-day equipment and methods the distance of the Sun is known extraordinarily accurately but it is fun to attempt the original method. Calculations can be a little tricky but Internet web sites available provided a simple screen to fill in contact times and instantly receive a Sun–Earth distance from your times. One site required also the exact coordinates of longitude and latitude of the observation point (the author carried a GPS receiver for this purpose). Using two web sites, distances of 138.56 and 134 million kilometers were obtained – a reasonable

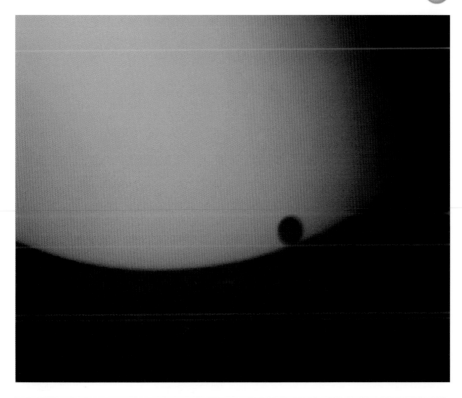

Figure 15.20.

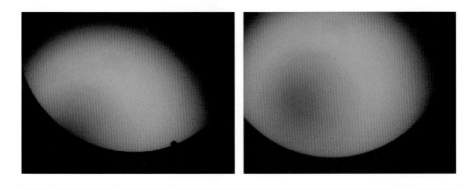

Figure 15.21.

result considering the error between official times and those recorded was only a few seconds. The Sun has a mean distance from the Earth of 149,597,893 km.

And finally, a gallery of snapshots of three methods of indirect projection of the Sun's image (figures 15.22 to 15.24).

Figure 15.22.

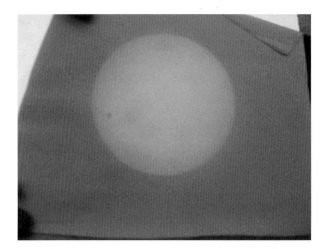

Figure 15.23.

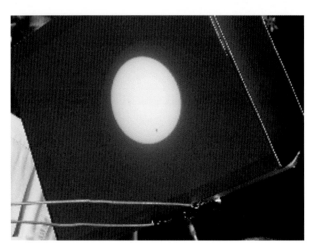

Figure 15.24.

Meticulous (but untidy) notes during the occasion of course (figure 15.25).

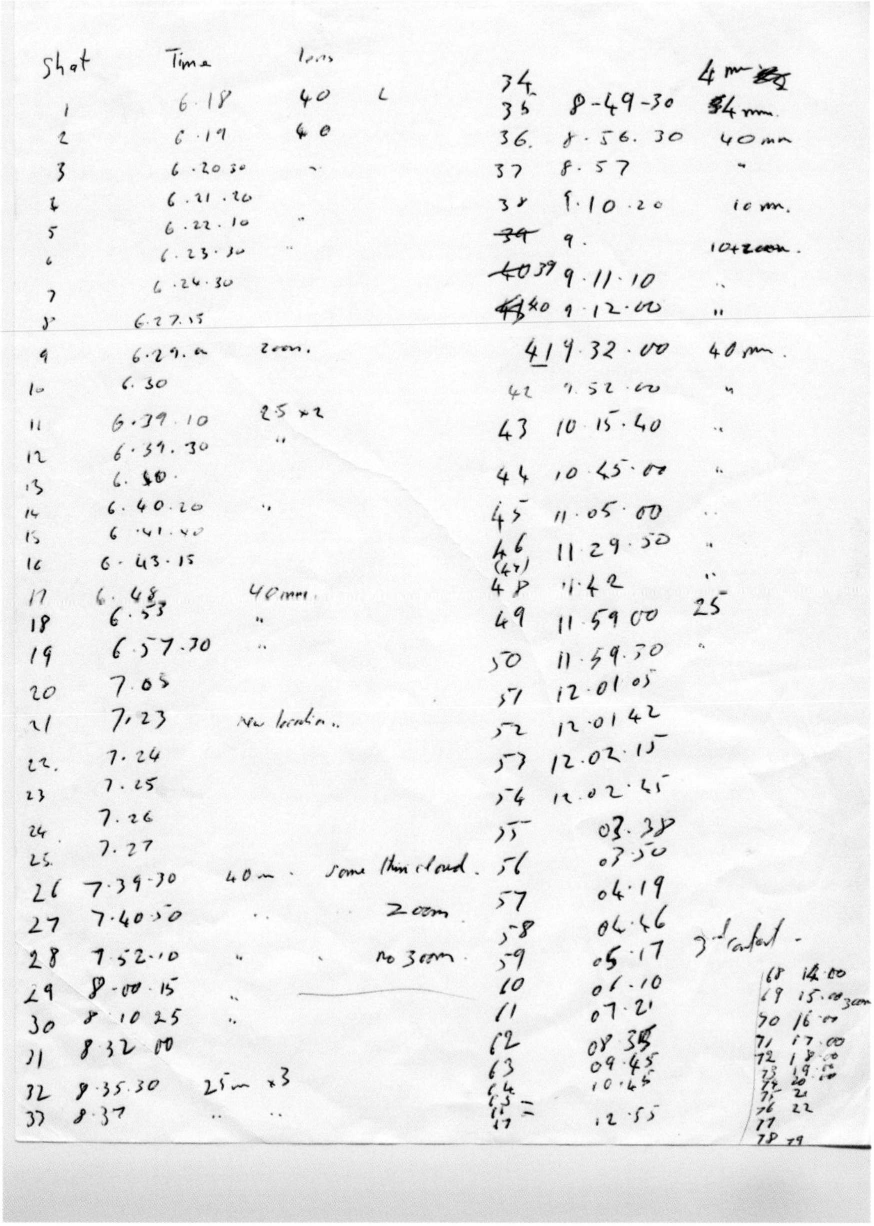

Shot	Time	lens	
1	6.18	40	L
2	6.19	40	
3	6.20 30	"	
4	6.21.20		
5	6.22.10	"	
6	6.23.30	"	
7	6.24.30		
8	6.27.15		
9	6.29.00	zoom.	
10	6.30		
11	6.39.10	25 ×2	
12	6.39.30	"	
13	6.40.		
14	6.40.20	"	
15	6.41.40		
16	6.43.15		
17	6.48	40mm.	
18	6.53	"	
19	6.57.30	"	
20	7.05		
21	7.23	New location..	
22.	7.24		
23	7.25		
24.	7.26		
25.	7.27		
26	7.39.30	40mm.	Some thin cloud.
27	7.40.30	..	Zoom.
28	7.52.10	"	No zoom.
29	8.00.15	"	
30	8.10 25	"	
31	8.32.00		
32	8.35.30	25mm ×3	
33	8.37	..	..

34		4mm ~~[illegible]~~
35	8-49-30	~~3~~4mm.
36.	8.56.30	40mm
37	8.57	"
38	9.10.20	10mm.
~~39~~	9.	10+zoom.
~~40~~ 39	9.11.10	"
~~41~~ 40	9.12.00	"
41	9.32.00	40mm.
42	9.52.00	"
43	10.15.40	..
44	10.45.00	"
45	11.05.00	..
46	11.29.50	"
(47)		
48	11.42	"
49	11.59 00	25
50	11.59.50	"
51	12.01.05	
52	12.01.42	
53	12.02.15	
54	12.02.45	
55	03.38	
56	03.50	
57	04.19	
58	04.46	
59	05.17	3rd contact.
60	06.10	
61	07.21	
62	08.38	
63	09.45	
64	10.45	
65 66 67 =	12.55	

68	14.00
69	15.00 zoom
70	16.00
71	17.00
72	18.00
73	19.00
74	20
75	21
76	22
77	
78	79

Figure 15.25.

CHAPTER SIXTEEN

And What Else?

Overview

The aim of this book is to show images that can be captured quickly, cheaply and easily in a limited area of a light-polluted night sky. To repeat the quote referred to earlier "stars and deep sky objects can be observed from a light-polluted back-yard, all you need is the time and the money to do it properly". Nevertheless, with this inexpensive equipment and technique very satisfying images can still be collected. There is also a great deal of difference between *observing* and *capturing* acceptable images. As objects appear in your area of sky under those rare excellent seeing conditions reasonable records of what you have seen can be produced. Here are some further examples.

Stars

The target is a group of stars known as the Beehive Cluster or Praesepe. It is also known as M44 from a list, compiled by the French Astronomer Charles Messier and published in 1781, of objects in the sky that could be confused with comets. M44 is to be found in the middle of the constellation of Cancer. It is an "open cluster", stars grouped together in no particular shape and somewhat spread out.

Once the cluster has been located and the camera positioned, apply the following settings: no zoom, no flash, shutter delay, brightness +2.0 and "night scene" or the equivalent settings on your particular camera. The image in figure 16.1 is as first seen on the computer following the usual download. Not much really until

Figure 16.1.

Figure 16.2.

the bright spots are emphasized using the brightness and contrast adjustments. The beauty of the cluster can now be seen (figure 16.2) and is quite a good record of what was visualized through the telescope.

Nebulae

Many satisfactory images of star groups, constellations and easy nebulae can be obtained using this technique although patience is required to wait for really good seeing conditions in a light-polluted environment. Shown in figure 16.3 is a completely unprocessed single image of the Orion nebula followed by a touch of brightness and contrast control (figure 16.4). With more patience, enhancement and stacking significantly improved images are available.

Comets

Comets are notoriously unpredictable as far as their brightness is concerned. They range from stunning displays visible to the naked eye, such as Hale–Bopp (1997), to smudges that pass with very little recognition. However, chances

Figure 16.3.

Figure 16.4.

Figure 16.5.

mustn't be missed even of capturing just enough of an image to prove you witnessed it. It is certainly good practice for a more prominent visitation later.

The picture in figure 16.5 was taken of comet Machholz as it exited Taurus during January 2005. It is the tiny smudge above the bright spot of a star. Big deal! But! By combining half a dozen images in RegiStax, as above, it can be made into a recognizable comet feature (figure 16.6).

Figure 16.6.

CHAPTER SEVENTEEN

In Conclusion

Remember the following:

Take as many pictures during a session as possible. Chance will play a great part in capturing stunning or unusual images.

The greatest of care must be taken with focusing. Even in poor seeing conditions a well-focused image can be worthy of a place in your collection if the subject is unusual or part of an event. Refocus frequently to ensure best adjustment.

Figure 17.1.

The images shown were all obtained without efforts to reduce light pollution or heat radiation from the ground. The telescope was rarely equilibrated for temperature unless by chance – photography was frequently commenced within 5–10 minutes. The "magic ingredient", the camera cradle, was constructed quickly and simply from wood, bits and cheap tripod available. No prototype, mark 1, 2 or 3, just one construction whatever it looked like. Just get on with it. Capture lots of amazing pictures and learn the theory and how to improve later.

Become thoroughly used to your camera, its facilities and operation. Practise in the comfort of your home to ensure complete familiarity with flicking through menus and settings.

Experiment as much as possible. You only need one "magic" picture out of the 100 or so to be pleased with the night's work.

Don't forget that your camera is also for general use on its own. Many sky images without the telescope are fascinating or picturesque (figure 17.1).

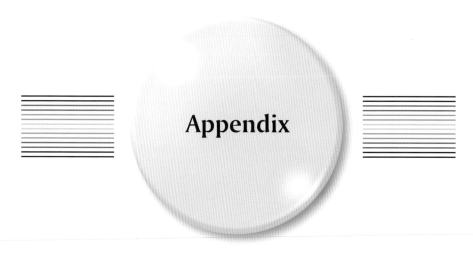

Appendix

1 Binoculars

For those aspiring astrophotographers who are hesitant to purchase a telescope mention was made of a DIY platform to support a camera to capture images through binoculars. The few illustrations given here should enable one to be constructed from easily obtainable parts (figures A.1 and A.2).

The platform is a piece of tongue and groove floorboard wood 30 cm × 15 cm. Saw it in two approximately 12 cm from one end (figure A.3). This part takes the camera. Add a door bolt to serve as a safety device to prevent accidental tipping of the camera support (figure A.4). Drill holes to take appropriate threaded bolts to secure the camera on one section and binocular clamp on the other. The clamp must have a screw thread to allow

Figure A.1.

Figure A.2.

Figure A.3.

Figure A.4.

Figure A.5.

attachment to a camera tripod. Affix two hinges to join the two sections together (figure A.5). A pencil of hexagonal cross section pushed into the groove locks the two sections together firmly. When the camera section is hinged down it is possible to view the sky object for focusing.

2 Helpful References

Drop in to Tony's website – www.givemethemoon.com for brand new galleries, assistance and advice for your own adapter, news, features and your chance to post your own findings.

There are many books to assist the amateur astrophotographer both in technique and identification of features recorded.

I suggest the following three are adequate in the first instance to help in the identification of features and facts for images captured and for further understanding of some of the astronomical principles involved.

The Data Book of Astronomy, Patrick Moore, Institute of Physics Publishing, 2000.

The Hatfield Photographic Lunar Atlas, ed. Jeremy Cook, Springer-Verlag,, 1999.

Atlas of the Universe, Patrick Moore, Philips, 1999.

And to assist with choice of telescope:

Astronomy with a Budget Telescope, Patrick Moore and John Watson, Springer-Verlag, 2003.

And for more details of the Moon and Sun:

Patrick Moore on the Moon, Cassell & Co., 2001.

The Cambridge Encyclopedia of the Sun, Kenneth R. Lang, Cambridge University Press, 2001.

Moonwatch, Peter Grego, Philips, 2003. The Philips package including this book also comes with a very useful and detailed Moon map.

The Modern Moon, Charles A. Wood, Sky Publishing Corporation, 2003.

Atlas of the Moon, A.Rukl (ed.T.W.Rackham), Hamlyn, London, 1990.

And to alert to astronomical events:

Yearbook of Astronomy, ed, Patrick Moore, Macmillan. Published annually.

3 For Further Progress

Do join your local astronomical society. It is the greatest source of information as well as providing much support, encouragement and social enjoyment with like-minded people.

Do read your telescope and camera manual.

Do attend astronomy exhibitions to keep abreast of what is available and to look out for bargains on the day.

Do experiment with equipment and conditions to produce even better pictures.

4 Glossary

Albedo	The proportion of light falling on a non-luminous body that is reflected from it. The Moon's albedo is 7%.
Apogee	The furthest point in an elliptical orbit from the center of the Earth.
Barycenter	The center of mass of a system of two bodies. The barycenter of the Earth–Moon system lies 1707 km below the Earth's surface.
Celestial Sphere	An imaginary sphere of indefinite size surrounding the Earth with Earth at the center.
Declination (dec)	The equivalent of Earth's latitude superimposed on the sky. Telescopes are manually or motor driven to follow latitude movement – the height above the observer's horizon. A clamp may be provided to hold a particular setting during viewing.
Earthshine	A slight illumination of the part of the Moon in shadow when it is a thin crescent.
Ecliptic	The apparent path of the Sun against the star background over the course of a year.
Ejecta	Material propelled outwards from an explosive event such as that produced by bombardment of the Moon to form craters.
Eyepiece	A lens or combination of lenses used to magnify the image formed by the telescope. The higher the number the lower the magnification. A 25 mm eyepiece produces a larger image than a 40 mm. Barlow lenses are added to the eyepiece to multiply the magnification by, typically, two or three. Hence, ×2 or ×3 Barlow.
Ghost crater	A heavily eroded or submerged crater such that only a small portion can be seen above the surrounding landscape to enable appreciation of its location.
Libration of the Moon	The wobble of the Moon. The main directions of libration are up–down and sideways which allow viewing of slightly more than 50% (in fact 59%) of the Moon's surface over a period of time.
Perigee	The nearest point in an elliptical orbit from the center of the Earth.

Pixelated	Reduction in resolution of an image to the point where individual pixels (picture elements) become apparent to produce a blotchy picture.
Rays or ray system	Bright streaks or splash effect resulting from debris thrown out from impact craters.
Reflector telescope	A telescope that uses a concave mirror to focus light as its main feature. A refractor uses a lens.
Rill(e)s or Rimæ	Cracks or long narrow lines of small craters on the Moon.
Star diagonal	Added to the rear of a telescope to direct the light at right angles to assist viewing. Eyepieces are placed in the star diagonal.
Terminator	The line between the sunlit and dark portion of the Moon (or planet, etc.).
Transient Lunar Phenomena. TLPs	Short-lived events observed on the surface of the Moon.

5 Index to Moon Features – Craters Unless Otherwise Stated

Sizes given in this list will contain discrepancies and anomalies that inevitably follow from the often wide variation of sizes given in texts researched.

Feature	Size (km)	Page
Abulfeda	65	172
Aestuum, Sinus	230	57, 123, 127
Agarum, Cape		69
Aggripa	44	113
Agricola, Montes	160	140
Airy	37	89
Albategnius	114	88, 89, 107, 110, 114, 175
Alexander	82	67
Aliacensis	80	88
Almanon	49	172
Alpes, Montes	261	67, 105
Alpes, Vallis	166 × 20	105
Alpetragius	40	110, 111, 175
Alphonsus	108	60, 107, 108, 110, 150, 177
Altai, Rupes	427	89, 106, 172
Ammonius	8	108
Ampère, Mons	30	76
Anaxagoras	50	102
Ancient Newton or Bliss		78, 100
Angström	9	80
Anville	16	71
Apenninus, Montes	401	75, 120
Apianus	63	88, 175

Feature	Size (km)	Page
Arago	26	73, 118, 161, 171
Archimedes	83	75, 78, 153, 168
Archytas	31	105
Argaeus, Mons	50	129
Argelander	34	89
Ariadaeus, Rima	220	112
Ariadaeus	11	112
Aristarchus	40	58, 80, 119, 137, 190
Aristillus	55	168
Aristoteles	87	66, 67, 153
Arzachel	96	60, 88, 107, 110
Atlas	87	74, 150
Australe, Mare	900	56
Autolycus	39	168
Babbage	143	103
Bailly	300	143, 144, 147
Baily	27	143
Ball	41	83, 11
Bancroft	13	75
Bartels	55	131
Bayer	47	87
Beer	10	75
Bernouilli	47	72
Berosus	74	124
Bessarion	10	80, 140
Bessel	15	63, 64, 123
Bettinus	71	88, 142, 143, 149
Bianchini	38	79
Billy	45	135, 174
Birmingham	98	100
Birt	16	110, 157, 177
Birt, Rima	50	157
Blagg	5	175
Blancanus	105	88, 148
Bliss (Ancient Newton)		78, 100
Bode	19	168
Boscovich	46	113
Brayley	15	80
Brianchon	145	103
Bruce	7	175
Bullialdus	61	149, 177
Burkhardt	57	72
Cajal	9	171
Campanus	48	150
Capella	49	127
Capuanus	59	116, 150
Cardanus	50	131, 190

Feature	Size (km)	Page
Carmichael	20	70
Carpetus, Montes	400	120, 181
Carrel	16	73, 161
Cassini	57	67, 105
Catharina	104	60, 92, 105, 106, 107
Caucasus, Montes	550	105, 157
Cavalerius	64	132
Cavendish	56	174
Cayley	14	113
Cepheus	39	74
Clavius	245	111, 147, 148, 149
Clavius C	22	148
Clavius D	27	148
Clavius J	10	148
Clavius N	11	148
Cleomedes	126	71
Cobra's Head	10	137
Cognitum, Mare	340	177
Colombo	76	91, 92
Copernicus	107	47, 58, 75, 95, 116, 119, 120, 171, 172, 181, 194
Cordillera, Montes	1,500	132
Cremona	85	103
Crisium, Mare	570	56, 58, 60, 67, 69, 70, 71, 125, 179, 180
Crüger	46	134
Cuvier	75	149
Cyrillus	98	60, 92, 105, 106, 107
Daguerre	40	106
Darwin	130	135
Davy	35	82, 108
Davy A		108
De Gasparis	30	86
De La Rue	134	74
Delisle	25	79, 157
Delisle, Mons	30	79
Dembowski	26	114
De Morgan	10	113
Desargues	85	103
Descartes	48	172
Deslandres	256	83, 85, 110, 111
Dionysius	18	113
Diophantus	17	79, 157
Dolland	11	172
Donati	36	89
Doppelmayer	64	86, 149
Draper	8	75

Feature	Size (km)	Page
Eddington	125	131
Egede	37	67, 105
Endymion	123	74, 125
Epidemiarum, Palis	300	57, 116, 150
Epigenes	55	102
Eratosthenes	58	75, 116, 123
Eudoxus	67	66, 67, 153
Euler	27	80
Fabricius	78	92
Faraday	70	149
Fauth	12	120
Faye	37	89
Fecunditatis, Mare	840	47, 56, 71, 72, ch 9.15, 126
Feuillée	9	75
Flamsteed, Ghost Crater	21	173
Fontenelle	38	102
Fracastorius	124	106
Fra Mauro	101	82, 171, 172
Franklin	56	74
Frigoris, Mare	1,350	56, ch 9.2, 76, 100, 103, 105, 127, 144, 181
Gardner	18	129
Gassendi	110	86, 149, 157, 174
Gassendi A	33	85
Gassendi B	25	85
Gauss	177	124, 125
Geminus	85	72
Goclenius	72	91, 92
Godin	35	113, 114
Goldschmidt	125	102
Greaves	14	125
Grimaldi	222	58, 60, 131, 132, 134, 178, 190
Gruithuisen	16	79, 157
Gruithuisen, Mons	20	79
Guericke	63	82, 172
Gutenberg	74	92
Hadley, Mons	25	169
Hadley, Rima	80	112, 169
Haemus, Montes	560	63, 64, 123
Hahn	84	124
Hainzel	70	88, 146, 150
Halley	36	172
Hansteen	45	135, 136, 172
Harbinger, Montes	90	80, 140
Hedin	143	132
Heinrich	6	75

Feature	Size (km)	Page
Heinsius	64	85, 146
Heis	14	79
Helicon	24	95
Hell	33	83, 85, 110
Heraclitus	90	149
Hercules	69	74, 150
Herodotus	34	80, 137
Herschel	41	110
Herschel, C.	13	79, 80
Herschel, J.	165	190
Hesiodus, Rima	300	116
Hevelius	118	132
Hind	29	172
Hipparchus	138	107, 110, 172
Hommel	126	89
Hooke	37	74
Horrocks	30	110
Huggins	65	149
Humboldt	207	126
Humboldtianum, Mare	160	56, 125
Humorum, Mare	400	56, ch 9.11, 149
Huxley	4	75
Hyginus, Rima	220	113, 114, 116, 175
Hypatia	40	73
Hypatia, Rima	206	171
Imbrium, Mare	1,300	56, 59, 64, 67, ch 9.5, 80, 82, 95, 153, 157
Iridum, Sinus	200	57, 78, 95, 181
Isidorus	42	127
Jansen	23	171
Janssen	199	92, 93, 150
Julius Caesar	90	64, 113
Jura, Montes	380	67, 79, 80, 81, 95
Kepler	32	58, 78, 80, 116, 119, 140, 172
Kirch	11	78, 100
Kircher	73	88, 142, 149
Krafft	51	131, 190
Krieger	22	80
Kundt	10	82
Lade	56	114
Lagalla	85	146
Lalande	24	82
Lambert	30	75, 80
Lamèch	13	66
Lamont	106	73, 170
Landsteiner	6	78, 95

Feature	Size (km)	Page
Langrenus	127	92
Laplace, Promontorium		80
Lavinium, Promontorium		70
Leakey	12	127
Lee	41	86
Le Verrier	20	95
Lexell	63	83
Licetus	75	149
Lick	34	70, 125
Liebig	37	86
Liebig, Rupes	180	86
Linné	2	95, 157
Lister, Dorsa	290	123
Lockyer	34	92
Lohrmann	31	132
Longomontanus	145	85, 146
Lubiniezky	43	177
Lunicus, Sinus, Bay of Lunik	100	168
Maclear	20	73, 116, 171
Macrobius	64	72
Mädler	28	106
Magelhaens	41	92
Maginus	194	148, 153
Mairan	40	81
Manilius	39	113
Marginis, Mare	360	56, 125
Marius	58	140
Maurolycus	114	150
McDonald	7	95
Medii, Sinus	350	57, 175
Mee	132	88
Menelaus	26	58, 63, 64
Mercator	46	150
Mersenius	84	60, 136, 157, 174
Messala	125	74
Milichius	12	194
Miller	61	149
Mitchell	30	66
Moltke	6	170
Montanari	76	146
Moretus	114	148, 153
Mortis	150	57
Mösting	25	82, 123
Mouchez	81	102
Murchison	57	114, 168
Nasireddin	52	149
Nasmyth	76	87, 142

Feature	Size (km)	Page
Nebularum, Palas	150	57
Nectaris, Mare	333	56, 72, 74, 89, ch 9.15, 105, 106, 127
Neper	137	125
Nielson	10	80
Nubium, Mare	750	56, ch 9.9, 116, 150, 157, 173, 174
Olivium, Promontorium		69
Orientale, Mare	327	56, 69, 132, 135
Orontius	105	148
Palisa	35	82
Pallas	46	114, 168
Peirce	19	70
Philolaus	70	102
Phocylides	114	87, 142
Piazzi Smyth	13	78, 100
Picard	23	70, 123
Piccolomini	87	89, 106
Pico, Mons	25	77, 78, 100
Pitatus	97	157
Pitiscus	82	89
Piton, Mons	25	78, 105
Planitia Descensus		178
Plato	100	59, 74, 95, 96, 98, 100, 102, 190
Playfair	48	88, 89
Plinius	43	73, 123, 171
Porter	52	147
Procellarum, Oceanus	2,568	56, 67, ch 9.7, 119, 137, 157, 171, 178, 181
Proclus	28	58, 67, 129
Promontorium Heraclides		181
Protagoras	22	105
Posidonius	95	123
Ptolemaeus	164	60, 107, 108, 110
Purbach	115	60, 107, 110
Putredinis, Palus	180	57
Pyrenaeus, Montes	250	92
Pythagoras	142	102
Pytheas	20	75
Recta, Rupes (The Straight Wall)	134	110, 177
Regiomontanus	129	60, 107, 110, 111
Reiner	30	140
Rheita, Vallis	445	93
Riccioli	146	132
Riphaeus, Montes	189	86, 171
Ritter	29	73, 116, 171

Feature	Size (km)	Page
Robinson	24	190
Rømer	40	131
Roris, Sinus	500	57
Rosenberger	95	93
Ross	24	73, 116, 171
Rost	49	87
Rümker, Mons	70	140, 141
Rupes Altai	480	89, 106, 172
Rupes Recta		110
Russell	103	131
Rutherford	50	147
Sabine	30	73, 116, 171
Sasserides	90	85
Scheiner	110	87, 148
Schiaparelli	24	178
Schickard	227	87, 142, 144, 146
Schiller	179 × 70	86, 146, 149
Schröter	35	83, 123
Schröter Valley (Vallis Schröteri)	200	78, ch 9.8, 81, 83, 123, 137, 178, 181
Segner	67	143
Seleucas	43	131, 178
Serenitatis, Mare	650	56, 58, ch 9.1, 67, 72, 73, 95, 123, 127, 129, 157, 173
Sharp	39	79
Sirsalis	42	135
Sirsalis, Rima	400	135
Smirnov, Dorsa	130	123
Smithson	6	71
Smythii, Mare	360	56, 125
Sömmering	28	83
Somnii, Palus	240	57, 70
Somniorum	230	57
South Massif		173
Spumans, Mare	139	56
Stadius	69	120
Stag's Horn Mountains		110
Stiborius	44	89
Stöfler	126	149
Strabo	55	74
Straight Wall	134	110, 177
Struve	170	131
Sulpicius Gallus	12	64
Swift	11	70
Taruntius	56	71, 72
Taurus–Littrow		173
Teneriffe, Montes	110	78, 100

Feature	Size (km)	Page
Thales	31	74
Theaetetus	25	67
Theophilus	110	60, 92, 105, 106
Timocharis	33	75
Torricelli	22	74
Tralles	43	72
Tranquillitatis, Mare	873	47, 56, ch 9.3, 72, 106, 116, 123, 127, 161, 170, 173
Triesnecker	26	114
Triesnecker, Rima	200	175
Tycho	85	2, 5, 47, 57, 58, 64, ch 9.10, 89, 111, 119, 144, 149, 153, 175, 177, 190
Ukert	23	175
Undarum, Mare	243	56, 125
Van Biesbroeck	9	80
Vaporum, Mare	245	56, 63, 64, 72, 113, 127, 175
Vinogradov	11	120
Vitello	42	86
Vitruvius	29	129
Vlacq	89	92, 150
Vogel	27	89
Walter	128	60, 88, 107, 110
Wargentin	84	87, 142, 144
Werner	70	88, 89
Whewell	14	113
Wilhelm	106	146
Wilson	69	142
Wöhler	27	90
Yerkes	36	70
Zucchius	64	88, 142, 143, 144, 149
Zupus	38	135, 136

Index

Note: This index contains only the most well-known lunar features. For a complete listing, see the Index to Moon Features on p. 259.

Able probes, 167
albedo
—of Mars's moons, 198
—of Moon, 48
Aldrin, Buzz, 170
Ancient Newton, 78, 100
aperture for telescope, 4, 5
Apollo missions, 72, 74, 112, 167–173
—photo of landing sites, 169
Arago domes, 73, 116, 161, 171
Ariel, 198–201
Armstrong, Neil, 170
astronomy, references on, 257
AstroStack, 212, 217
Atlas, 198

"baby" (lunar feature), 157
Baily, Francis, 143
Barlow lens, 10, 23–24, 210, 219
barycenter, 48
batteries (camera), 9
Bean, Alan, 171
Beehive Cluster, 247
Big Bang, 31–32
Big Crunch, 33
Big Whack, 44–45
binoculars, 5
—camera platform for, 255–257
—overview of Moon features visible with, 55–61
birds, unexpected sightings of, 192
Blue Moon, 51
brightness setting, 8
—for planets, 214, 219
—for stars, 247
—See also exposure

C8. See Celestron C8 telescope
Callisto, 198–201, 210
camera, 6–9
—batteries, 9
—brightness setting, 8, 214, 219, 247
—exposure (see exposure)
—flash, 8
—image quality (see image quality)
—memory, 7
—monitor (rear viewing screen), 8, 26–27

camera (continued)
—"night scene" setting, 247
—panorama, 9
—picture storage and retrieval, 9
—pixels, 7, 24–26
—shutter delay/self-timer, 7–8, 22, 201, 211, 214, 247
—simulated ISO selection, 25–26
—spot metering (see spot metering)
—taking movies with, 211–212, 216–217
—use of (see photographic techniques)
—zoom, 8–9, 215, 219, 247
camera holder, 254
—and binoculars, 255–257
—instructions for building, 13–19
—use of, 21–22
Cassini mission, 198, 207
Celestron C8 telescope, 10–11
—camera-to-telescope holder instructions, 13–19
Cernan, Eugene, 173
Charon, 199–201
"claw" (lunar feature), 150
Clementine probe, 168
"Cobra's Head" (moon feature), 137
collimation, 208–209
Collins, Michael, 170
color
—and atmospheric conditions, 28–29, 75, 116, 182
—and lunar eclipses, 182
—lunar surface variations, 29, 75, 87, 95, 142, 161
—optical effects, 190
comets, photographic techniques for, 249–251
computer models of Moon origin, 46
computer techniques
—combining single images, 213–214, 217–218, 249, 251
—crop, zoom, and flip functions, 27
—enhancements of Earthshine, 188–190
—enlargement of photographs, 105, 106, 116, 190, 238–239
—processing movies for planetary photography, 211–212, 216–217

Conrad, Charles, 171
Cook, James, 242
Copernicus
—characteristics of, 47
—as naked eye target, 58
—ray system, 119
—See also Index to Moon Features on p. 259
cosmic microwave background, 32
cosmology, 31–34
craters
—buried/destroyed, 73, 78, 83, 100, 110, 120, 161, 170
—concentric, 71, 75, 86
—origins of, 47
—See also ray systems; terracing; and specific craters listed in the Index to Moon Features on p. 259

Darwin, George, 43
Data Book of Astronomy (Moore), 198
Davis, D. R., 44
Deimos, 197–201
density variation in lunar surface, 73
dew shield, 29
Dione, 199–201
domes, lunar
—Arago domes, 73, 116, 161, 171
—Gardner Megadome, 129
—Marius region, 140
—Mons Rümker, 140–141
—Oceanus Procellarum region, 80
—southeast region, 90
Doppler shift, 32–33
Duke, Charles, 172

Earth
—composition of, 46
—ocean tides, 45
—and origins of the Moon, 43–46
—tilt of axis, 46
Earthshine, 188–190
eclipses
—lunar, 181–188
—solar, 143, 228–230
enlargement of photographs, 105, 106, 116, 190, 238–239
erosion, absence of on Moon, 47

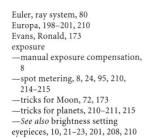

Euler, ray system, 80
Europa, 198–201, 210
Evans, Ronald, 173
exposure
—manual exposure compensation, 8
—spot metering, 8, 24, 95, 210, 214–215
—tricks for Moon, 72, 173
—tricks for planets, 210–211, 215
—*See also* brightness setting
eyepieces, 10, 21–23, 201, 208, 210

faculae, 228, 233
fault lines, lunar, 105, 113. *See also* rills
film speed, 25–26
finder scope, 40
First Quarter Moon, 50, 52, 53
Fisher, Osmond, 43
flash, 8
focus, 21–22, 26–27, 210, 253
foreshortened images, 69, 131
Full Moon, 51, 52, 53
—lunar eclipses, 181–188
—northeast region viewed at, 74
—and shadows, 126
—Tycho region viewed at, 83, 146

galaxies, 32–33
Galileo Galilei, 43, 46, 219
Ganymede, 198–201, 210
Gardner Megadome, 129
"ghost" craters, 78, 120, 170. *See also* Ancient Newton
Gibbous Moon, 50, 51, 53
Gold, Thomas, 178
Gordon, Richard, 171
graben, 105
gravity, 32, 33, 222
Great Red Spot, 210, 214
Grimaldi
—creation of, 47
—location of, 60–61
—as naked eye target, 58
—*See also* Index to Moon Features on p. 259
Gruithuisen, Franz von Paula, 79

Hagomoro probe, 168
Hale–Bopp (comet), 249
Halley, Edmund, 242
Hartman, W. K., 44
Hatfield Lunar Atlas (Cook, ed.), 63
"Helmet" (lunar feature), 157
Hiten satellite, 168
Hubble expansion, 33
Huygens probe, 198, 203, 207
hydrogen-alpha filter, 223, 238

Iapetus, 199–201
image quality, 24–26
—color variation due to atmospheric conditions, 28–29, 75, 116, 182

image quality *(continued)*
—combining single images, 213–214, 217–218, 249, 251
—dew shield, 29
—and focus, 26–27
—and low angle observations of the Moon, 40, 150
—and movie processing, 212
—seeing conditions (*see* seeing conditions)
—selection of, 7
—and temperature equilibrium, 27
Io, 198–201, 210
Irwin, Jim, 169
ISO, simulated selection of, 25–26

Japan, Hagomoro probe, 168
Jupiter
—characteristics of, 209–211
—moons of, 198–203, 210
—photographic techniques for, 209–214

Kepler
—as naked eye target, 58
—ray system, 78, 80, 119
—*See also* Index to Moon Features on p. 259
Kuiper belt, 199

Last Quarter Moon, 51, 52, 53
lava domes. *See* domes, lunar
lava plains, 78, 131
—Mare Frigoris region, 66–67
—Mare Imbrium region, 78
—Mare Nubium region, 82
—Mare Tranquillitatis region, 69, 71
—Oceanus Procellarum region, 79
—southwest region, 87
lava wrinkles, 110, 157
—Mare Crisium region, 71, 125
—Mare Imbrium region, 168
—Mare Serenitatis region, 123
—Mare Tranquillitatis region, 72, 73, 161, 170
libration, 40, 102, 125, 131, 146, 149
light pollution, 247, 254
limb, features foreshortened at, 69, 131
Luna probes, 167, 168, 172, 178–181
lunar eclipses, 181–188
lunar missions, 167–181. *See also* specific missions and probes
lunation, 51

M44, 247
Machholz (comet), 251
magic ingredient. *See* camera holder
manual exposure compensation, 8
Mare Crisium, 67–69
—foreshortened image, 69
—lava wrinkles, 71, 125
—location of, 60–61

Mare Crisium *(continued)*
—and lunar missions, 179–180
—as naked eye target, 58
—*See also* Index to Moon Features on p. 259
Mare Fecunditatis, 47, 91–92. *See also* Index to Moon Features on p. 259
Mare Frigoris, 66–67. *See also* Index to Moon Features on p. 259
Mare Humorum, 85–86. *See also* Index to Moon Features on p. 259
Mare Imbrium, 75–78
—lava plains, 78
—lava wrinkles, 168
—and lunar missions, 168, 181
—origins of, 47, 64, 80
—ray systems, 75
—*See also* Index to Moon Features on p. 259
Mare Nectaris, 91–92. *See also* Index to Moon Features on p. 259
Mare Nubium, 82–83, 173. *See also* Index to Moon Features on p. 259
Mare Serenitatis, 63–65
—lava wrinkles, 123
—and lunar missions, 168–170
—ray systems, 63–64
—*See also* Index to Moon Features on p. 259
Mare Tranquillitatis, 67–74
—and Apollo moon landing, 72, 170
—creation of, 47
—lava plains, 69, 71
—lava wrinkles, 72, 73, 161, 170
—as naked eye target, 58
—ray systems, 67, 69, 70, 129, 131
—*See also* Index to Moon Features on p. 259
maria
—list of, 56–57(table)
—overview of targets, 55–59
—*See also* specific maria in the Index to Moon Features on p. 259
Mariner 10 probe, 232
Mars, moons of, 197–201
mascon, 73
Mattingly, Thomas, 172
Mead ETX70AT, 11
Melosh, Jay, 46
memory, 7
Menelaus
—as naked eye target, 58
—ray system, 63–64
Mercury
—characteristics of, 232–233
—lack of moons, 197
—transits, 231–233
Messier, Charles, 247
metering. *See* spot metering
microwave background radiation, 32
Miranda, 199–201
missions to the Moon. *See* lunar missions
Mitchell, Edgar, 172
Monarch of the Moon, 119
monitor (rear viewing screen), 8, 26–27

Moon
—absence of erosion, 47
—albedo of, 48
—appearance and position of, 39–40, 48–53
—barycenter, 48
—color variation due to atmospheric conditions, 28–29, 75, 116, 182
—color variation due to surface characteristics, 29, 75, 87, 95, 142, 161
—compared to other Solar System moons, 199, 200, 201
—composition of, 44–46
—distance to Earth, 194
—early orbit, 45, 49
—Earthshine, 188–190
—far side of, 57
—features by region (see Index to Moon Features on p. 259; Moon features)
—libration ("wobble"), 40, 102, 125, 131, 146, 149
—lunar eclipses, 181–188
—lunar shadow (umbra), 229
—lunation, 51
—mascon, 73
—missions to (see lunar missions)
—naked eye features (see naked eye features of the Moon)
—orbital path, 49
—orbital period, 49
—origins of, 43–46
—overview of targets, 55–61
—phases, 49–53
—photographic techniques, 39–40, 95–165
—references on, 257
—relative position of Moon and Sun in the sky, 52–53
—shadows, 70, 96–98, 150, 153
—soil composition, 175
—surface characteristics, 46–48, 73, 178
—terminator, 53, 66
—transient phenomena, 86, 190–192
Moon features
—Mare Frigoris region, 66–67
—Mare Humorum and south Oceanus Procellarum region, 85–86
—Mare Imbrium region, 75–78
—Mare Nubium region, 82–83
—Mare Serenitatis region, 63–65
—Mare Tranquillitatis region, 67–74
—Maria Fecunditatis and Nectaris region, 91–92
—northeast region, 74
—northwest region, 80–81
—Oceanus Procellarum region, 78–80
—overview, 55–61
—southeast central region, 88–89
—southeast region, 89–90, 92–94
—southwest region, 86–88
—Tycho region, 83–85

Moon features (continued)
—See also craters; domes, lunar; fault lines, lunar; lava plains; lava wrinkles; ray systems; rills; and Index to Moon Features on p. 259
moons, Solar System, 197–206, 210
Moore, Sir Patrick, 78, 198, 234
MOV format, 212, 217
movies, processing, 211–212, 216–217

naked eye features of the Moon, 55–61
—craters, 57–59, 95, 146
—Earthshine, 188–190
—location of features, 59–61
—maria, 58, 69
—ray systems, 58, 83
nebulae, photographic techniques for, 249
Neptune, moons of, 198–201
Nereid, 199
neutrons, 32
New Moon, 50, 51, 52, 53
—Earthshine, 190
—Mare Crisium viewed at, 69
—and solar eclipses, 228
"night scene" setting, 247
nucleosynthesis, 32

Oberon, 199–201
Oceanus Procellarum, 78 80, 85 86
—and lunar missions, 171–172
—ray systems, 78, 80, 119
—See also Index to Moon Features on p. 259
O'Neill, John, 69
open clusters, 247–249
orbit of Moon, 45, 49
Orbiter probes, 167
orientation of pictures, 22–23, 27
Orion nebula, 249
Oscillatory Universe model, 33
Otford, Kent, scale model of Solar System, 34–37

Pan, 198
Pandora, 198
panorama, 9
phases of the Moon, 49–53
phases of Venus, 219
Phobos, 197–201
photographic techniques, 21–29, 214, 247
—and binoculars, 255–257
—brightness setting, 214, 219, 247
—cautions about daylight photography, 24, 40–41, 223
—and color (see color)
—combining single images, 213–214, 217–218, 249, 251
—for comets, 249–251
—computer enlargements, 105, 106, 116, 190, 238–239
—and Earthshine, 188–190

photographic techniques (continued)
—exposure tricks, 72, 173, 210–211, 215
—filming near the terminator, 66
—focus, 26–27, 253
—identifying unexpected images, 192–195
—image quality (see image quality)
—for Jupiter's moons, 200–203
—and lunar eclipses, 181–188
—magic ingredient (see camera holder)
—for Moon features, 95–165
—for nebulae, 249
—orientation of pictures, 22–23, 27
—overview of targets, 39–42, 55–61
—placement issues for set-up, 27
—for planets, 207–220
—practice techniques, 21–26, 254
—processing movies, 211–212, 216–217
—resolution of photographs, 24–25
—for Saturn's moons, 203–206
—simulated film speed selection, 25–26
—spot metering, 8, 24, 95, 210, 214–215
—for stars, 247–249
—for Sun, 223–230 (see also transits)
—taking many photographs in one session, 113, 190, 253
—and temperature equilibrium, 27
—and transient lunar phenomena, 190–192
—for transits, 231–245
—unexpected images, 190, 192–195
photons, 32
picture storage and retrieval, 9
Pioneer probes, 167
pixels, 7, 24–26. See also enlargement of photographs
placement issues for set-up, 27
planets, 208
—formation of, 33–34
—Jupiter, 209–214
—moons of, 197–206
—photographic techniques, 41–42, 207–220
—Saturn, 214–219
—transit of Mercury, 231–233
—transit of Venus, 231, 234–245
—transits, 231–245
—Venus, 219–220
Plato
—creation of, 47
—as naked target, 59, 95
—See also Index to Moon Features on p. 259
Pluto, moon of, 199–201
practice techniques, 21–26, 254
Praesepe, 247
Proclus
—as naked eye target, 58
—ray system, 67, 69, 70, 129, 131
—See also Index to Moon Features on p. 259
Prometheus, 198

prominences, 222, 238
Prospector probe, 168
protons, 32

Quaoar, 199
quarks, 32

Ranger probes, 167, 170, 175, 177
ray systems
—Copernicus, 119
—Euler, 80
—Kepler, 58, 78, 80, 119
—Menelaus, 58, 63–64
—as naked eye features, 58, 83
—Proclus, 67, 69, 70, 129, 131
—Timocharis, 75
—Tycho, 83, 144, 146, 149
redshift, 32–33
reflectivity of the Moon, 48
RegiStax, 212, 217, 251
resolution of photographs, 24–25
Rhea, 199–201
rills
—Alphonsus, 177
—Goclenius and Gutenberg, 92
—Grimaldi area, 131
—Hadley, 112, 169
—Rima Ariadaeus, 112, 113
—Rima Birt, 157
—Rima Hesiodus, 116
—Rima Hyginus, 113, 114, 116
—Rima Hypatia, 116, 171
—Sirsalis, 135
—Triesnecker area, 114, 116
—See also fault lines, lunar; Index
  to Moon Features on p. 259
Roche, Edouard, 44
Roosa, Stuart, 171
Rupes Recta (Straight Wall;
  Sword), 110, 157, 177

satellites, unexpected sightings of,
  192–195
Saturn
—moons of, 198–199, 203–206
—photographic techniques, 214–219
—rings of, 214, 218
Schmidt–Cassegrain telescope,
  10–11, 208
Schmitt, Harrison, 173
Scott, David, 169
Sedna, 199
seeing conditions, 66, 102
—and low angle observations of
  the Moon, 40, 150
—and observations of deep space
  objects, 247, 249
—and transient lunar phenomena,
  190
shadows
—lunar shadow (umbra), 229
—on the Moon, 70, 96–98, 150, 153
—and Saturn's rings, 218
Shepard, Alan, 172
shutter delay/self-timer, 7–8, 22,
  201, 211, 214, 247
"skull" (lunar feature), 157

Smoothed Particle Hydrodynamics
  (SPH), 46
software. See AstroStack; RegiStax
solar eclipse, 143, 228–230
solar filters, 10–11, 40, 223, 238–239
Solar System
—formation of, 33–34
—moons, 197–206
—origins of the Moon, 43–46
—scale model at Otford, Kent, 34–37
—transits, 231–245
—See also Sun; specific planets and
  moons
Soviet Union, 167
space race, 167
SPH. See Smoothed Particle
  Hydrodynamics
spot metering, 8, 24, 95, 210, 214–215
stars, photographic techniques for,
  247–249
storage and retrieval of pictures, 9
Straight Wall. See Rupes Recta
Sun, 221–230
—cautions about photographing,
  11, 24, 40–41, 223
—characteristics of, 221–223
—faculae, 228, 233
—filters, 10–11, 40, 223, 238
—origin of, 34
—photographic techniques, 40–41
—prominences, 222, 238
—relative position of Moon and
  Sun in the sky, 52–53
—solar eclipse, 143, 228–230
—sunspots, 221, 223–228
—transits, 231–245
sunspots, 221, 223–228
Surveyor probes, 167, 170, 171, 173,
  174–177
Sword. See Rupes Recta

telescopes
—aperture, 4, 5
—Barlow lens, 10, 23–24, 210, 219
—camera-to-telescope holder
  instructions, 13–19
—Celestron C8, 10–11, 13–19
—collimation, 208–209
—dew shield, 29
—eyepieces, 10, 21–23, 201, 208,
  210
—Mead ETX70AT, 11
—orientation of pictures, 22–23
—placement issues, 27
—selection of, 9–11
—solar filters, 10–11, 40, 223,
  238–239
—temperature equilibrium, 27
—See also photographic techniques
temperature equilibrium, 27
terminator, 53, 66. See also Moon
  features
terracing
—Aliacensis, 88
—Apianus, 89
—Aristoteles, 66
—Bayer, 87
—Bullialdus, 149
—Hansteen, 174

terracing (continued)
—Mersenius, 174
—Piccolomini, 89
—Plinius, 73
—Rømer, 131
—Tycho, 146
—Werner, 89
—Zucchius and Bettinus, 88
—See also further page numbers for
  these craters in the Index to
  Moon Features on p. 259
Tethys, 199–201
Theia, 46
Thomas, David, 34
tides, ocean, 45, 49
Timocharis, ray system, 75
Titan, 198–201, 203
Titania, 198–201
transient phenomena (TLPs), 86,
  190–192
transits, 231–245
—Mercury, 231–233
—Venus, 231, 233–245
tripod, camera, 15, 17–18
Triton, 198–201
Tycho, 83–85
—characteristics of, 47, 83, 146
—and lunar missions, 175–177
—as naked eye target, 57–58, 146
—ray system, 83, 144, 146, 149
—See also Index to Moon Features
  on p. 259

Umbriel, 199–201
universe. See cosmology
Uranus, moons of, 198–201

Venus
—characteristics of, 234
—lack of moons, 197
—phases of, 219
—photographic techniques, 219–220
—transits, 234–245
viewing screen. See monitor (rear
  viewing screen)
Voyager 2, 198

waning crescent Moon, 51
waning gibbous Moon, 51
waxing crescent Moon, 50
waxing gibbous Moon, 50
webcam movies, 211
website, author's, 257
Wilkins, Percy, 70
"wobble" of Moon. See libration
Wood, Charles, 129
Worden, Alfred, 169
wrinkles. See lava wrinkles

Young, John, 172

Zond probes, 167
zoom settings, 8–9
—for planets, 215, 219
—for stars, 247